# Heredity
## &
# Environment

# Heredity
# & Environment

*edited by*
*A. H. Halsey*

**The Free Press**
*A Division of Macmillan Publishing Co., Inc.*
NEW YORK

The Free Press
A Division of Macmillan Publishing Co., Inc.
866 Third Avenue, New York, N.Y. 10022

First American Edition 1977

Jan 22 '79

Library of Congress Catalog Card Number: 77-2530

ISBN 0-02-913670-9

Printed in Great Britain

printing number

1   2   3   4   5   6   7   8   9   10

# Contents

# Acknowledgements

The editor and publisher wish to thank the following for permission to reproduce material from the publications listed below:

Professor J. M. Thoday and the Eugenics Society for an extract from 'Geneticism and Environmentalism' in *Biological Aspects of Social Problems* (1965); the executors of the estate of Professor Th. Dobzhansky and Oxford University Press for 'Natural Selection in Mankind', in *The Structure of Population*, edited by G. A. Harrison and A. J. Boyce (1972); the Eugenics Society for 'Sociology, Biology and Population' by Dr A. H. Halsey in *Eugenics Review*, Vol. 59 (1967); Professor N. Tinbergen and the Royal Society of London for extracts from *Proceedings of the Royal Society of London*, B 182, 385–408 (1972); Professor V. C. Wynne-Edwards and the Rockefeller University Press for extracts from 'Population Control and Social Selection in Animals', in *Genetics*, edited by D. V. Glass (1968); G. Hawthorn and the Open University for 'Some economic explanations of fertility', based on Open University talk 'Making sense of society', reference JPM 950H212; Professor Mary Douglas and Routledge and Kegan Paul Ltd. for 'Population Control in Primitive Groups' in *British Journal of Sociology*, Vol. XVII, No. 3, September 1966; B. Benjamin and the Eugenics Society for 'Social and Economic Differentials in Fertility', in *Genetic and Environmental Factors in Human Ability* (1965); Professor J. E. Meade and the British Academy for 'The Inheritance of Inequalities', in *Proceedings of the British Academy*, Vol. LIX (1973); B. K. Eckland and Grune and Stratton Inc. for 'Social Class Structure and the Genetic Basis of Intelligence', from *Intelligence: Genetic and Environmental Influence*, edited by R. Cancro (1971); Dr A. H. Halsey and Routledge and Kegan Paul Ltd. for 'Genetics, Social Structure and Intelligence', *British Journal of Sociology* (1958); M. Young and the British Psychological Society for 'In search of an explanation of social mobility', in *British Journal of Statistical Psychology*, Vol. XVI, Part I, May 1963; Professor A. R. Jensen for his article 'Race and Mental Ability', from *Racial Variation in Man*, Institute of Biology Symposium No. 22 (London: Institute of Biology/Blackwell, 1975); Professor P. V. Tobias and the *Journal of Behavioural Science* for 'IQ and the Nature-Nurture Controversy', *JBS*, 2 (1), 1974. G. Smith and T. James and Carfax Publishing Co. for 'The effects of pre-school education: some American and British evidence', from *Oxford Review of Education*, Vol. I, No. 3, 1975; Professor W. F. Bodmer and the Times Newspapers Ltd. for 'By the colour of their genes', *The Times Literary Supplement*, 3 August 1973.

# Introduction

## A. H. Halsey

The difference of natural talent in different men is, in reality, much less than we are aware of; and the very different genius which appears to distinguish men of different professions, when grown up to maturity, is not upon many occasions so much the cause, as the effect of the division of labour. The difference between the most dissimilar characters, between the philosopher and the common street porter, for example, seems to arise not so much from nature, as from habit, custom, and education.
(Adam Smith, *An Enquiry Into The Nature and Causes of The Wealth of Nations,* 1793)

But the truth is, that though human institutions appear to be the obvious and obstructive causes of much mischief to mankind; yet, in reality, they are light and superficial, they are more feathers that float on the surface, in comparison with those deeper seated causes of impurity that corrupt the spring, and render turbid the whole stream of human life.
(Thomas R. Malthus, *An Essay On The Principle of Population,* 1798)

Respected authorities would persuade us to opposite conclusions. The task now is to set out the method and point the direction which will lead to a secure understanding of heredity and environment. Accordingly, the body of this book consists of readings on methodological and substantive issues chosen to illustrate the approach and the achievement of the human sciences in theory and research concerning the ancient problem of nature and nurture.[1] Five points may, however, be argued in order to justify the selection and arrangement of the readings.

The first argument, fundamental to the book as a whole, is that relevant theory relies on the idea of *interactive causation* and is *integrative* between the biological and social sciences. Second, it is essential to recognize fully the fact of genetic complexity in man.

Third, and of corresponding importance, is the complexity of environmental variation in human cultures which requires for its appreciation a developed sociological perspective. Fourth must be noted the frequency with which problems of social policy assume theories or raise questions about the causal relations between biological and social inheritance. The contemporary problem of transmitted deprivation is a case in point and is examined below with reference to Jencks's recent and widely discussed work on inequality in America (Jencks 1972). Finally, and consequentially, a moral responsibility can be inferred and must be asserted as intrinsic to scientific inquiry into what is literally a matter of life and death.

## 1 Interaction and integration

As a first step towards the understanding of continuity and change in human nature the biological and historical records permit us an initial and firm generalization. Human nature is modifiable and is indeed constantly being modified both genetically and phenotypically. The difficulty is to say how much. To what extent are traits, in Tinbergen's phrase, 'environment resistant'? The point here – and that of the *Introduction* to this book – is to abandon as unscientific any conception of human nature as immutably fixed or unlimitedly plastic. Truth lies in neither of these extremes. Yet neither is it to be found in the compromising middle. We have to learn to ask different and more difficult questions. The nature–nurture controversy has to be broken down to refer to what characteristics, under what circumstances, in what periods of time. We shall see and illustrate in the following chapters that a pure geneticism or a pure environmentalism in the search for explanations of human behaviour are both crude fallacies. Professor Thoday, in Chapter 1, clears the ground by destroying these extremist simplicities and he goes further towards disabusing the beginner of widely circulating errors in explaining that 'no characteristic is largely acquired'. In fact *all* characteristics are *both* acquired *and* genetic. Genotypes determine potentialities. Environment determines which or how much of these potentialities should be realized in living human beings. In other words, the common mark of all viable theories on the nature–nurture issue are theories of *interaction*. All formulations which polarize nature from nurture are essentially misleading. Heritability is partially determined by the range of variation of the environment in which it operates, and conversely the strength of environmental determination is in part a function of the genetic variability of the population living in the environment in question. Thus, for example, to say that intelligence is 80 per cent attributable to genetic inheritance is a strictly meaningless proposition, at least until it is specifically applied to a defined population in time and space. The student of the human sciences has to learn from the outset to think in terms of interactive causation.

This, then, is the basic mode of thought which informs the modern human science approach to heredity and environment. Contributions to it are to be found in all of the biological, psychological and social sciences. I have tried, therefore, in selecting a short set of introductory readings, to take examples of writings from the relevant organized academic subjects. Thus, Thoday, Dobzhansky, Gibson and Bodmer are geneticists; Tinbergen and Wynne-Edwards are ethologists; Tobias is a biologist, Benjamin a demographer, Meade an economist; Hawthorn, Eckland, Young, Smith and James are sociologists, Jensen a psychologist, and Douglas an anthropologist. The relations between these academic fraternities, however, are not and have not been conspicuously harmonious ones (see Chapter 3 below for a brief discussion of the estrangement of the biological and social sciences; for a fuller history see Haller 1963). Polarization between geneticism and environmentalism continues to obstruct collaboration across disciplinary boundaries, but the work of such writers as Dobzhansky, Tinbergen, G. H. Harrison (Harrison and Boyce 1972) or Torsten Husén (1974) exemplify integrative tendencies which promise much for the solution of the problems of heredity and environment.[2]

The modern human sciences are seen at their best when under the influence of this integrative impulse, and I have tried to choose examples of this type of approach which illustrate scientific progress. As will be seen, these are most obviously to be found in recent work on fertility and population control, on intelligence and social stratification, and on the explanation of differential achievement between races. Accordingly, after Part I, which elaborates the approach advocated in this *Introduction*, the chapters are arranged in three further parts devoted to these areas of substantive research.[3]

## 2 The complexity of inheritance

The ramifications of interactive method are not immediately obvious. They involve appreciation of three principal complexities – that of the genetic composition of man, that of the range of possible environmental variation, and finally that of the processes of interaction themselves. After millions of years we are bequeathed from these sources an evolved, rich and bewildering human inheritance.

Perhaps the most important message of modern genetics, with its growing knowledge of polymorphism, is the awesome scale of genetic variability within the single species of *homo sapiens*. To think of a single or monozygotic determination of any human trait is to concentrate on the rare. And of all the billions of individual people who have ever existed, monozygotic twins apart, it is probable to the point of near certainty that no two have ever had the same genetic composition (see below, p. 189). Similarly, it is all too easy to underestimate the actual and even more the possible variations in human environment – in which the biological inheritance is allowed phenotypic realization – unless adequate attention is paid to the anthropological and historical record. In consequence, when the interaction of these vast complexities is considered, the danger is always again that of oversimplification and the challenge is to discover the crucial principles of interplay.

There are no easy short cuts. When we use such a word as inheritance we refer to the outcome of a long and intricate historical process of genetic and environmental interaction. The present stage of imperfect knowledge invites search for economical simplifications. Robert Ardrey's recent *The Hunting Hypothesis* (Ardrey 1976) is an example, aptly subtitled as 'a personal conclusion concerning the evolutionary nature of man'. His attempt to specify the major elements of human inheritance has the merit that he adopts an interactionist approach. He avoids a simple geneticism and stresses, for example, the importance of the effects of climatic change on the course of hominid evolution. Nevertheless, his hunting hypothesis – which gives central place to the inheritance by *sapiens* of those characteristics of *homo australopithecus* which evolved from a long history of carnivorous living by hunting of large mammals in groups and with tools – remains inspired guesswork. Such theorizing is to be welcomed as offering guidance to further research but carries with it the danger of pointing to false trails. In his review of Ardrey, Geoffrey Gorer points out that:

> Africa abounds with rats and mice and other small deer, with molluscs and crustacea and fishes and reptiles, all of them edible and only the crustacea leaving fossils, and all, or nearly all, of them easily 'gathered' by women and children. But if females could feed themselves and their children, even if not with the choicest meats, then the role of the male hunter is comparatively diminished; and *The Hunting Hypothesis* demands that it be aggrandized as much as possible until the Neolithic revolution and the invention of agriculture, when grain replaced game. (*New Statesman and Nation,* 16 July 1976)

## Biological and cultural evolution

Evolution and natural selection cannot be separated from the problem of heredity and environment. Professor Dobzhansky in Chapter 2 makes it clear that both genetic and environmental forces are constantly at work in the control of the direction of change in human characteristics and human behaviour. It is of massive importance in explaining continuities and variations in the life of *homo sapiens* to recognize the cumulative power of cultural evolution. Again our thinking has to be interactive and to recognize the continued operation of both biological and cultural evolution in their complex interplay with each other. A full understanding of this interplay is what takes us beyond Darwinian evolutionary theory to the modern human sciences. To say this is specifically not to espouse the fallacy that cultural intervention by man in effect brought biological evolution to a halt. To speak of the cumulative power of cultural evolution is once again to emphasize the interactive process, behind which the 'Darwinian' forces of natural selection operating on the random variations of genetic individuality continue to operate. Culture continuously changes the conditions under which these forces act. For example, the mating and breeding of early man took place under cultural conditions which permitted only relatively small gene pools in comparative isolation. Such conditions maximized 'experiment' in natural selection, promoted genetic drift, homozygosity and the genetic differentiation of tribal groups. As cultural changes brought increasing power over nature, so the size of populations and gene pools increased and internal differentiation (i.e. increasing complexity of the division of labour) permitted changes in patterns of mating, and rates of fertility and mortality. Looked at in this way culture can be thought of as a burgeoning source of historical innovation in human behaviour.

At the same time a developed appreciation of culturally determined variation and change need be in no way inconsistent with the conception of contemporary human behaviour as rooted in the genetic characteristics of early man. Professor Tinbergen takes this as his starting point in Chapter 4 and stresses the 'conviction that cultural evolution has not been random, but that it has affected principally those aspects of ancestral behaviour equipment that were relying most on individual programming, and has changed to a far lesser extent the more internally programmed, the more resistant trait, which therefore must be taken to reflect most clearly our ancestral heritage. Evidence of such environment-resistant traits can be expected to come mainly from two sources. Whatever is least variable between cultures, and whatever is least variable within a culture and appears even in spite of environmental pressures in a culturally modified society, is most likely to reveal an ancient, environment-resistant 'deeper structure'. It is only when this deep structure is laid bare that we can address ourselves intelligently to the possibilities of deliberately creating optimal environments for mankind and deciding on the validity of those theories which assert that the conditions of modern life put excessive strain on human adaptability, i.e. are in conflict with what Tinbergen calls the deeper structure of our biological inheritance.

### 3 The complexity of environment
The human scientist has to learn to be sociological as well as interactionist in his thinking. In the long period after the explosion of Spencerian social Darwinism in the nineteenth century it became a cliché of sociological thought that in the understanding of human behaviour 'sociology begins where biology leaves off'. Interactionism exposes the falsity of this doctrine but the need for sociological sophistication remains if environmental variation is to be taken fully into account. Professor Thoday, writing as a geneticist, again points to this necessity in Chapter 1 when he tells us that we can say nothing about intelligence as measured *except in a relation to a population*.

To say that human scientists must think sociologically is not to decry interest in the individual organism or in observation of the individual. It is, however, to assert the superficially nonsensical point that the sum of individuals is greater than the parts. We shall encounter the need for this type of sociological or supra-individual conceptualization throughout the book. Thus, for example, we have to learn to recognize not the individual but the local population or deme as the unit of evolution, i.e. the unit of change in the interactive balance of nature and nurture. Geoffrey Hawthorn and Mary Douglas illustrate this point in their essays (Chapters 6 and 7) on the group norms which influence fertility. And we shall eventually come across a paradox of the human sciences in the problem formulated by J. B. S. Haldane as 'genes for altruism'. Wynne-Edwards's Chapter 5, 'Population Control and Social Selection in Animals', argues the case for recognition, despite the Darwinian tradition of emphasis on individual competition for survival, for a theory of intergroup as distinct from individual selection in the evolution of species and the evolution of human behaviour.

## 4 Social policy and human science

Modern cultural evolution contains purposive elements. Modern man tries deliberately to make himself by planning systematic changes in his own environment.

Direct control of genetic inheritance is, of course, familiar to farmers and has always been possible in the case of human populations. Moreover, the sophistication of methods of genetic control is a likely outcome of advances in genetic science. Meanwhile, given what we already know about the interactive character of behaviour, environmental control is also of con-siderable potential power for the same purposes. And it is power for both good and ill. There are, in other words, inescapable moral and political aspects of the sciences of man. Hence it is now widely if belatedly recognized that intelligent social policy requires new relations between the human sciences and govern-ment.

To say this is not to advocate either a technocracy or a servile technology. All of the conflicting interests of science and politics which are traditionally familiar remain, indeed become more urgent, as political debate needs increasingly to be informed by scientific knowledge. Governmental decisions concerning taxation, family allowances, food subsidies, housing densities, location of hospitals, number of university places, and so on, will all influence the genetic composition of our descendants and, conversely, the genetic composition of the population influences the significance of governmental decisions. Politics and science are accordingly related – the only question is how intelligently. There is a large set of challenges here which takes us beyond the possible compass of this book. Nevertheless, their existence must be fully grasped in any introduction to the human sciences. For this reason we shall turn at the end of this *Introduction* to consider a particular problem of social policy – that of 'transmitted deprivation' – which brings out the characteristic way in which political discussion in contemporary society raises the nature–nurture issue.

## 5 Scientific and social responsibility

Our discussion below of the case of transmitted deprivation also serves to remind us that there is nothing especially new about the passionate political sentiments which overhang discussion of heredity and environment. To explain human behaviour is *ipso facto* to indicate what is politically possible and often to suggest what is socially desirable. Theories and policies imply each other. Human science and human politics are accordingly intertwined and this was always so. The intrinsic moral dimension of the problem of nature and nurture has to be recognized and accepted by responsible scientists. In this sense science is also political activity and nowhere less than in the realm of the genetic and environmental causes of human action. It is, of course, the business of science to seek truth, and this is our primary aim in this book. Without truth there can be no science. Science can serve social justice, but it can also serve malevolence. Hence it is a dangerous weapon. Where, as here, we are dealing with matters of race and class and the distribution of opportunities and rewards, the scientists must address the student or the layman with the highest possible standards of cautious precision, recognizing the task as a political act with responsibilities not only to science but also to society.

The intercourse of common men does not habitually accept the discipline of responsibility either to science or to an enlightened social ethic. On the side of prejudice it must be observed that vulgar language is soaked in ethnocentrism and class discrimination. The *Oxford Dictionary* faithfully reflects widespread usage in nouns like 'nigger' and definitions of 'Jew' or 'to welsh'. On the side of scientific inquiry it may equally be noted that, for example, 'it is no use flying in the face of nature' is a traditional folk-catholic theory of the inevitable causes of human behaviour. Common language abounds in such theories: 'You can't make a silk purse out of a sow's ear', 'Like father like son', 'Blood is thicker than water'. A convenient store of handy proverbs is available in human culture to suit particular occasions, thus affording a convenience which science eschews. Popular sayings are not asked, collectively, to face the test of consistency and they are characteristically used to underline confirming examples rather than to undergo trials by falsification. Accordingly, the storehouse of customary anecdote also contains contrary theories: 'If at first you don't succeed, try, try again', 'Where there is a will there is a way', 'Fortune favours the brave', and so on.

Uncommon literature reflects the popular ragbag albeit more epigrammatically. 'Manners makyth man', the Wykehamist motto, asserts in effect that silk purses can be made out of sows' ears. *Man Makes Himself* is the title of Gordon Child's famous book on prehistory and is not obviously consistent with the Marxist pronouncement that 'It is not man's consciousness which determines his existence but his existence which determines his consciousness'.

To introduce order and precision into the body of answers to the question of what is inevitable and what modifiable in human behaviour is a primary task of the human sciences. This book is an attempt to introduce the reader to the modes of thought, the reasoning and the evidence which hold out the possibility of transforming traditional gossip into systematic science, with respect to a fundamental set of questions about human life. Unhappily, scientists and students have not always observed the full rigour of their scientific and social responsibilities.

The recent revelations concerning the work of Sir Cyril Burt are a sad case in point.[4] The effect of reappraisal of Burt's interpretation of his twin studies is to strengthen the case for reducing the degree of heritability to be attributed to intelligence. The ultimate scientific implication of discarding Burt's empirical work on intelligence is probably negligible. Nevertheless, the harm done by him, whether by motivated carelessness or deliberate dishonesty, is serious. To some greater or lesser degree he distorted opinion and policy in matters affecting the lives of people and he betrayed the trust which enables scientists to bring rationality to bear on the conduct of human affairs.

Another example is provided by the half-educated thugs who physically assaulted Professor Eysenck and prevented him from lecturing at the London School of Economics in 1973, thus denying the scientific and social ethic which justifies academic freedom. But Professor Eysenck, if more politely, is also open to criticism. His *Race, Intelligence and Education* (1971), which appeared in Britain in the wake of the storm raised in America by the writings of Jensen, Herrnstein and Shockley, is an example of failure to observe the high standards of scientific and social responsibility to which they no doubt aspire. Eysenck must have known that the English language is soaked in racism: yet he chose to discuss the logic of experiment by reference to a tasteless and offensive limerick about Miss Starkie who had an affair with a Darkie.

Sometimes he has seemed less than scrupulous in his representation of other scholars. He attacked those who took Jensen to be saying that 'scientific evidence proved Negroes to be inferior to whites in intelligence ...' In fact, Eysenck (1971, p. 17) tells us, Jensen 'does not suggest any of these things, either directly or by inference'. But look at what Jensen actually wrote: 'All the major facts would seem to be comprehended quite well by the hypothesis that something between one-half and three-fourths of the average IQ difference between American Negroes and whites is attributable to genetic factors ...' (Jensen 1973a). Does this or does it not suggest to a reasonable lay reader that scientific evidence proves Negroes to be genetically inferior to whites in intelligence?

Much of Eysenck's science is lucidly argued. But it is more aggressive than truthful with respect to sociology. His account of the nature–nurture controversy is incomplete. Sociologists undifferentiated are flayed for unscientific environmentalism, but the excesses of social Darwinism and of racist biological theories, against which many sociologists have fought, are not discussed.

Eysenck says: 'Sociology, on the whole, has not observed the lesson of science, that knowledge cannot be acquired by leaving out of account alternative hypotheses and concentrating on those which appeal to the research worker's prejudices'. Thus, what might otherwise have been a clear and fair discussion is marred at many points by a limited knowledge of sociology. One example must suffice. 'In contradiction to human conditions', he writes, 'the rats were assigned to conditions; they had no chance to select their preferred environment. This is an important difference.' One can have little confidence in the sociological knowledge about race of a man who dismisses constraints on human social freedom so lightly.

The central fact to be explained and with which Eysenck is concerned in this book is that Negroes in America have measured intelligence (IQ) on average 10 to 15 points below that of American whites. Does this mean that the black races are naturally inferior to the white races in their intellectual make up? In Part IV below there are chapters by Jensen, Tobias, Bodmer, and Smith and James which address themselves to this issue. At this introductory point we can make the question more precise, if narrower, by speaking only of American Negroes and American whites and not assuming that each is a representative genetic sample of any wider population which we might want to call a race. Second, by the word 'naturally' we can mean 'genetically determined'. The question is then whether the average difference is (*a*) genetic, (*b*) a combination of genetic and environmental influences, or (*c*) environmental. Eysenck shows his bias and wastes time by calling the second type of theory hereditarian and the third type environmentalist. Obviously the first is hereditarian, the third environmentalist, and the second a combination of the two. And we can say on the evidence that the hereditarian hypothesis is definitely false.

How do we decide between the other two hypotheses? First, the evidence from biologically related and unrelated people reared apart and together makes it scientifically clear that both heredity and environment cause individual differences of measured intelligence. The problem of explaining IQ differences between racial groups therefore looks easy. If we know the amount of variance attributable to heredity and to environment, and if we know that the measured group differences are greater than the variation caused by environmental difference, then there must be a genetic component.

But unfortunately it is not as easy as that. The 4:1 ratio of heredity to environment used by Eysenck is a dubious oversimplification. The degree of difference in IQ which is attributable to heredity is itself partly determined by environment and *vice versa*. One environment may stimulate the development of a particular genetic potential and another may repress it, so that in the first situation heredity can have a greater influence than it could in the second. Neither genotypes nor environments are uniform in any population and the proportion of IQ variance attributable to either is in part a function of the variation of the other. As Professor J. M. Thoday (see p. 36 below) puts it:

> Unique genotype and unique environment are interacting in the development of each individual in unique ways, and, though we must classify individuals into groups for scientific, administrative or educational purposes, we ignore this uniqueness to our great loss and at our peril, and it makes nonsense of segregation of races justified solely on the basis of differences of average even if the average differences may be real.

Neither Jensen nor Eysenck would disagree with Thoday. Both, in fact, insist on the need to match individuals to individualized education. But in their concern to emphasize the importance of the genetic hypothesis against environmentalists they give the misleading impression that only hereditarians care about individuals. Science will serve individualism as well as itself and society better when both geneticism and environmentalism are dead.

## Jencks and the cycle of transmitted deprivation

To end this introduction we return to a widely discussed example of the contemporary problems which draw human science into the political arena. My object here is to illustrate in some detail how such problems may be approached in academic terms but with relevance to the political implications of scientific enquiry. The case in point is that of the theory of 'transmitted deprivation' which recently attracted the notice of a British Minister of Social Security, Sir Keith Joseph, with a consequent invitation to the Social Science Research Council to produce research findings.[5] And this particular problem can be usefully discussed in relation to Christopher Jencks's much debated book (1972) on a similar problem at the centre of American politics – the determinants of inequality of occupational status and income.

Christopher Jencks's *Inequality* is certainly relevant from the point of view of research on the transmission of deprivation and inequality, being motivated by a search for effective ways of reducing income inequality in the United States and being concerned, for the major part of its analysis, to debunk some conventional beliefs. In particular, he attacks the triple hypothesis that genetic inheritance, family background and schooling have much effect on inequalities of occupational status and income. In other words, Jencks produces, or more accurately reproduces from other sources, genetic and environmental evidence on the continuities and discontinuities of life chances for individuals in different generations. The more true Jencks's thesis that family background and schooling are *not* determinants of the occupational status and income of individuals, the less scope there is for a cycle of *transmitted* deprivation operating through families.

We shall have to return both to Jencks's conclusions and to the validity of his arguments with respect to these hypotheses. But before doing so we must examine what might be meant by the terms transmission, cycle, deprivation and inequality – all key words in the Social Science Research Council's programme of study inspired originally by Sir Keith Joseph's speech in June 1972 in which he suggested that deprivation often recurred in successive generations of the same family and called for explanation in order to promote effective preventive action.

As a matter of fact, genetics apart, the very concept of society implies transmission, given that men are mortals. If a society exists for longer than the lifetime of its current members there must be transmission. Six contexts of transmission can, however, be distinguished. First, we can imagine a society which undergoes no institutional change and in which there is exact reproduction in the sense that each individual marries once and once only and the married couple reproduces a son and a daughter who then go on to occupy the social position held by the parent of the same sex. In such a society transmission would be total, the cycle would be an exact intergenerational reproduction of social identities irrespective of genetic distributions; and if there were inequality and/or deprivation (whether defined with respect to the structure of the society as a whole or the position of individuals within it) these would continue indefinitely and unchanged. This lunatic first picture of a possible society is invented here in order to make it clear what could be meant by the strongest possible definition of socially transmitted deprivation and inequality. Moving to a second model, we can remove the assumption that individuals follow their parent of the same sex. Instead, we assume that they are allocated at random to the unchanged structure of social positions. In this second example it should be noticed that deprivation and inequality would exist to the same extent in each generation as in the original generation. But the cycle to which Sir Keith Joseph referred would scarcely exist. To be exact, the proportion of the population ($d$) who were caught in the cycle of deprivation in the sense of descending from ancestors who had themselves been deprived would be in the $n$th generation $d^n$. Thus if one-tenth of the original population were deprived, then one-thousandth of the generation of grandchildren would find themselves members of the deprivation cycle.

These two cases are represented in the middle column of Table 1. They assume a static social structure, i.e. one in which the total proportion of deprived members of society remains constant over the generations. But the first model assumes zero intergenerational mobility for individuals with the result that there is a static transmitted deprivation cycle equal to the total number of deprived persons while the second assumes the opposite extreme of 'perfect mobility' with the result that there is a contracting cycle of transmitted deprivation.

Table 1 *Typology of transmitted deprivation*

| Individual intergenerational mobility | Deprivation | | |
|---|---|---|---|
| | *Increasing* | *Static* | *Decreasing* |
| 0 | Expanding transmitted deprivation cycle | Castelike and static transmitted deprivation cycle | Contracting transmitted deprivation cycle |
| + | Indeterminate | Contracting transmitted deprivation cycle | Contracting transmitted deprivation cycle |

However, to be more realistic, we must remove the assumption of a static array of social positions. Theoretically, it is possible over the generations for inequality and/or deprivation to increase (through economic decline, regressive reallocation of life chances, population increase with static GNP, etc.). This case is not typical of the Western industrial world though it may approximate to some Third World conditions. It would, in any case, have two broad types represented in the first column of Table 1. In the first there would again be zero individual intergenerational mobility and hence an expanding transmitted deprivation cycle. In the second case, where there was intergenerational mobility the transmitted deprivation cycle would be indeterminate depending on whether individual mobility took more people out of the cycle than social impoverishment kept in.

Jencks's analysis of the United States has little or nothing to say about these third and fourth conditions, which in any case are not those which describe the recent history of Great Britain. For these we have to turn to a fifth and sixth alternative social situation in which there is growth in GNP *per capita* and, though this is arguable, a secular trend towards more equal distribution of life chances. Again, individual intergenerational mobility provides two subtypes as shown in the third column of Table 1. But in both the fifth and sixth case a contracting transmitted deprivation cycle would result. Under conditions of zero individual mobility the contraction of the cycle would be a simple function of the decline in the total amount of deprivation. In the other case, intergenerational mobility (provided that it involved the children of the deprived) would add to the rate of contraction of the cycle and this last case offers the easiest and fastest possibility of eliminating transmitted deprivation. It is also, incidentally, the case to which Britain approximates in the aspirations of its public policies. In other words, Britain is a country with an open class system, a history of economic growth and public policies aimed at redistributing wealth, income, health, education, housing, etc., in favour of the relatively disadvantaged.

This last (sixth) case also comes nearest to describing the USA which Jencks has analysed. In the generation before 1962 the USA experienced both economic growth and individual intergenerational mobility. The GNP *per capita* approximately doubled and the correlation between filial and parental status was about 0·4. Within this framework a closer, though still very imprecise, view of transmitted deprivation may be had by looking at the mobility tables in Blau and Duncan (1967), on which Jencks draws and from which the outflow percentages for all the low status categories which might be thought of as appropriately labelled among the deprived can be seen.[6] The percentages are such that a transmitted deprivation cycle continues to exist but probably accounts for a small minority of those who are to be counted as deprived on a definition applying to both the filial

Table 2 USA: Mobility from father's occupation to 1962 occupation, for males 25 to 64 years old: outflow percentages

| Father's occupation | Respondent's occupation in March 1962 | | | | | | | | | | | | | | | | | Total^a |
|---|---|---|---|---|---|---|---|---|---|---|---|---|---|---|---|---|---|---|
| | 1 | 2 | 3 | 4 | 5 | 6 | 7 | 8 | 9 | 10 | 11 | 12 | 13 | 14 | 15 | 16 | 17 | |
| *Professionals* | | | | | | | | | | | | | | | | | | |
| 1 Self-employed | 16·7 | 31·9 | 9·9 | 9·5 | 4·4 | 4·0 | 1·4 | 2·0 | 1·8 | 2·2 | 2·6 | 1·6 | 1·8 | ·4 | 2·2 | 2·0 | 0·8 | 100·0 |
| 2 Salaried | 3·3 | 31·9 | 12·9 | 5·9 | 4·8 | 7·6 | 1·7 | 3·8 | 4·4 | 1·0 | 6·9 | 5·2 | 3·4 | 1·0 | 0·6 | 0·8 | 0·2 | 100·0 |
| 3 Managers | 3·5 | 22·6 | 19·4 | 6·2 | 7·9 | 7·6 | 1·1 | 5·4 | 5·3 | 3·1 | 4·0 | 2·5 | 1·5 | 1·1 | 0·8 | 0·5 | 0·1 | 100·0 |
| 4 Salesmen, other | 4·1 | 17·6 | 21·2 | 13·0 | 9·3 | 5·3 | 3·5 | 2·8 | 5·4 | 1·9 | 2·6 | 3·7 | 1·7 | 0·0 | 0·8 | 1·0 | 0·3 | 100·0 |
| 5 Proprietors | 3·7 | 13·7 | 18·4 | 5·8 | 16·0 | 6·2 | 3·3 | 3·5 | 5·2 | 1·9 | 5·1 | 3·6 | 2·8 | 0·5 | 1·2 | 1·1 | 0·4 | 100·0 |
| 6 Clerical | 2·2 | 23·5 | 11·2 | 5·9 | 5·1 | 8·8 | 1·3 | 6·6 | 7·1 | 1·8 | 3·8 | 4·6 | 5·6 | 1·0 | 1·8 | 1·3 | 0·0 | 100·0 |
| 7 Salesmen, retail | 0·7 | 13·7 | 14·1 | 8·8 | 11·5 | 6·4 | 2·7 | 5·8 | 3·4 | 3·1 | 8·8 | 5·1 | 4·6 | 0·1 | 3·1 | 2·2 | 0·0 | 100·0 |
| *Craftsmen* | | | | | | | | | | | | | | | | | | |
| 8 Manufacturing | 1·0 | 14·9 | 8·5 | 2·4 | 6·2 | 6·1 | 1·7 | 15·3 | 6·4 | 4·4 | 10·9 | 6·2 | 4·6 | 1·7 | 2·4 | 0·4 | 0·1 | 100·0 |
| 9 Other | 0·9 | 11·1 | 9·2 | 3·9 | 6·5 | 7·6 | 1·5 | 7·8 | 12·2 | 4·4 | 8·2 | 9·2 | 4·6 | 1·2 | 2·8 | 0·9 | 0·3 | 100·0 |
| 10 Construction | 0·9 | 6·7 | 7·1 | 2·6 | 8·3 | 7·9 | 0·8 | 10·4 | 8·2 | 13·9 | 7·5 | 6·2 | 5·2 | 1·1 | 4·3 | 0·8 | 0·6 | 100·0 |
| *Operatives* | | | | | | | | | | | | | | | | | | |
| 11 Manufacturing | 1·0 | 8·6 | 5·3 | 2·7 | 5·6 | 6·0 | 1·4 | 12·2 | 7·3 | 3·2 | 17·9 | 6·9 | 5·1 | 4·0 | 3·5 | 0·8 | 0·6 | 100·0 |
| 12 Other | 0·6 | 11·5 | 5·1 | 2·5 | 6·6 | 6·3 | 1·4 | 7·1 | 9·3 | 4·9 | 10·4 | 12·5 | 5·9 | 2·1 | 4·2 | 0·9 | 1·1 | 100·0 |
| 13 Service | 0·8 | 8·8 | 7·4 | 3·5 | 6·0 | 9·0 | 1·9 | 8·0 | 6·4 | 5·4 | 11·7 | 8·1 | 10·5 | 2·7 | 3·3 | 1·0 | 0·2 | 100·0 |
| *Labourers* | | | | | | | | | | | | | | | | | | |
| 14 Manufacturing | 0·0 | 6·0 | 5·3 | 0·7 | 3·3 | 4·4 | 0·7 | 10·7 | 6·0 | 2·8 | 18·1 | 9·4 | 9·4 | 7·1 | 5·8 | 1·7 | 0·9 | 100·0 |
| 15 Other | 0·4 | 4·9 | 3·5 | 2·5 | 3·5 | 8·7 | 1·7 | 7·7 | 8·2 | 5·7 | 12·7 | 10·6 | 8·1 | 3·4 | 9·9 | 0·9 | 1·1 | 100·0 |
| 16 Farmers | 0·6 | 4·2 | 4·1 | 1·2 | 6·0 | 4·3 | 1·1 | 5·6 | 6·7 | 5·8 | 10·2 | 8·6 | 4·8 | 2·4 | 5·4 | 16·4 | 3·9 | 100·0 |
| 17 Farm labourers | 0·2 | 1·9 | 2·9 | 0·6 | 4·0 | 3·5 | 1·2 | 6·4 | 6·6 | 5·8 | 13·1 | 10·8 | 7·5 | 3·2 | 9·2 | 5·7 | 9·4 | 100·0 |
| 18 Total^a | 1·4 | 10·2 | 7·9 | 3·1 | 7·0 | 6·1 | 1·5 | 7·2 | 7·1 | 4·9 | 9·9 | 7·6 | 5·5 | 2·1 | 4·3 | 5·2 | 1·7 | 100·0 |

^a Rows as shown do not total 1000, since men not in the experienced civilian labour force are not shown separately.
^b Includes men not reporting father's occupation.
*Source:* P. Blau & O. D. Duncan, *The American Occupational Structure.*

Table 3 USA: Mobility from father's occupation in 1962 for males 25 to 64 years old: inflow percentages

| Father's occupation | Respondent's occupation in 1962 | | | | | | | | | | | | | | | | |
|---|---|---|---|---|---|---|---|---|---|---|---|---|---|---|---|---|---|
| | 1 | 2 | 3 | 4 | 5 | 6 | 7 | 8 | 9 | 10 | 11 | 12 | 13 | 14 | 15 | 16 | 17 |
| *Professionals* | | | | | | | | | | | | | | | | | |
| 1 Self-employed | 14.5 | 3.9 | 1.5 | 3.8 | 0.8 | 0.8 | 1.1 | 0.3 | 0.3 | 0.6 | 0.3 | 0.3 | 0.4 | 0.2 | 0.6 | 0.5 | 0.6 |
| 2 Salaried | 7.0 | 9.5 | 4.9 | 5.8 | 2.1 | 3.8 | 3.4 | 1.6 | 1.9 | 0.6 | 2.1 | 2.1 | 1.9 | 1.4 | 0.4 | 0.5 | 0.3 |
| 3 Managers | 8.7 | 7.9 | 8.7 | 7.0 | 4.0 | 4.4 | 2.6 | 2.7 | 2.6 | 2.2 | 1.4 | 1.2 | 1.0 | 1.8 | 0.7 | 0.3 | 0.3 |
| 4 Salesmen, other | 5.6 | 3.4 | 5.2 | 8.1 | 2.6 | 1.7 | 4.4 | 0.8 | 1.5 | 0.8 | 0.5 | 1.0 | 0.6 | 0.0 | 0.4 | 0.4 | 0.3 |
| 5 Proprietors | 18.5 | 9.6 | 16.5 | 13.2 | 16.3 | 7.1 | 15.2 | 3.5 | 5.2 | 5.7 | 3.7 | 3.4 | 3.7 | 1.6 | 2.0 | 1.5 | 1.6 |
| 6 Clerical | 4.9 | 7.3 | 4.4 | 5.9 | 2.3 | 4.5 | 2.6 | 2.9 | 3.1 | 1.2 | 1.2 | 1.9 | 3.2 | 1.5 | 1.3 | 0.8 | 0.0 |
| 7 Salesmen, retail | 0.9 | 2.3 | 3.0 | 4.7 | 2.8 | 1.8 | 2.9 | 1.4 | 0.8 | 1.1 | 1.5 | 1.1 | 1.4 | 0.1 | 1.2 | 0.7 | 0.0 |
| *Craftsmen* | | | | | | | | | | | | | | | | | |
| 8 Manufacturing | 3.8 | 8.3 | 6.1 | 4.3 | 5.1 | 5.7 | 6.3 | 12.0 | 5.1 | 5.1 | 6.2 | 4.7 | 4.8 | 4.5 | 3.2 | 0.5 | 0.4 |
| 9 Other | 4.0 | 7.0 | 7.4 | 7.9 | 6.0 | 8.0 | 6.1 | 6.9 | 11.0 | 5.8 | 5.3 | 7.8 | 5.4 | 3.8 | 4.1 | 1.2 | 1.2 |
| 10 Construction | 3.0 | 3.2 | 4.4 | 4.1 | 5.8 | 6.2 | 2.6 | 6.9 | 5.5 | 13.7 | 3.6 | 3.9 | 4.6 | 2.6 | 4.9 | 0.8 | 1.8 |
| *Operatives* | | | | | | | | | | | | | | | | | |
| 11 Manufacturing | 5.2 | 6.4 | 5.1 | 6.5 | 6.1 | 7.5 | 7.1 | 12.9 | 7.7 | 4.9 | 13.7 | 6.9 | 7.1 | 14.5 | 6.3 | 1.2 | 2.8 |
| 12 Other | 2.8 | 7.5 | 4.2 | 5.4 | 6.2 | 6.7 | 6.0 | 6.5 | 8.6 | 6.6 | 6.9 | 10.9 | 7.1 | 6.5 | 6.4 | 1.2 | 4.4 |
| 13 Service | 2.3 | 3.7 | 4.0 | 4.8 | 3.7 | 6.3 | 5.3 | 4.8 | 3.9 | 4.7 | 5.1 | 4.6 | 8.2 | 5.4 | 3.3 | 0.8 | 0.6 |
| *Labourers* | | | | | | | | | | | | | | | | | |
| 14 Manufacturing | 0.0 | 1.0 | 1.2 | 0.4 | 0.8 | 1.3 | 0.8 | 2.6 | 1.5 | 1.0 | 3.2 | 2.2 | 3.0 | 5.9 | 2.4 | 0.6 | 0.9 |
| 15 Other | 1.0 | 2.0 | 1.9 | 3.3 | 2.1 | 6.0 | 4.7 | 4.5 | 4.8 | 4.8 | 5.3 | 6.9 | 6.2 | 6.7 | 9.6 | 0.7 | 2.8 |
| 16 Farmers | 11.2 | 10.8 | 13.3 | 10.1 | 24.3 | 18.3 | 17.6 | 20.1 | 24.4 | 30.4 | 26.6 | 29.4 | 22.8 | 29.5 | 32.6 | 82.0 | 59.7 |
| 17 Farm labourers | 0.3 | 0.5 | 0.9 | 0.5 | 1.5 | 1.5 | 2.1 | 2.3 | 2.4 | 3.1 | 3.4 | 3.7 | 3.6 | 3.9 | 5.6 | 2.9 | 14.5 |
| 18 Total[a] | 100.0 | 100.0 | 100.0 | 100.0 | 100.0 | 100.0 | 100.0 | 100.0 | 100.0 | 100.0 | 100.0 | 100.0 | 100.0 | 100.0 | 100.0 | 100.0 | 100.0 |

[a] Columns as shown do not total 1000, since men not reporting father's occupation are now shown separately.

*Source:* P. Blau & O. D. Duncan, *The American Occupational Structure*, p. 39.

and parental generation. Occupational categories are, of course, at best a rough indicator of deprivation which needs a more refined definition to take into account such relevant factors as family size, single parentage, housing conditions and genetic attributes. Nevertheless, even if plural measures were exactly applied, the sixth type of small and contracting cycles seems most likely to emerge.

Looked at in inflow rather than outflow terms, i.e. the parental origins of those in low status categories, it appears from Table 3 that 9·6 per cent of 'other labourers' in 1962 were themselves the sons of 'other labourers'. From this inflow percentage the same inference may reasonably be drawn, i.e. a small and contracting cycle.

A slightly more sophisticated way of looking at the same data is to compare actual rates of mobility with the model of 'perfect mobility' which is defined by statistical independence of origins and destination. In the case of perfect mobility each destination group has the same distribution of origins as the total population, each origin group has the same distribution of destinations as the total population and all the indices of association are 1·0. Table 3 shows us the matrix of these indices. High self-recruitment between generations is represented by indices greater than 1. The index for 'other labourers' is 2·7, for labourers in manufacturing 3·8, and for farm labourers 4·1. These indices are not high and again the same inference may be drawn with respect to the scope of the cycle of transmitted deprivation.

Table 4 shows the matrix of indices of association for the British population in 1949 as studied by David Glass (1954) and his colleagues. It may be seen here that the highest level of self-recruitment is at the top of the society in Category 1 (professional and managerial families) but there is a relatively high level at the bottom of the society among semi-skilled and unskilled labourers. But, again, the figure is not so high as to suggest great scope for the transmitted deprivation cycle. It is 2·259. Comparable data on the British 1972 population are not yet published. But this new Oxford inquiry is likely to confirm the small scope for a transmitted deprivation cycle and to indicate contraction in the post-war generation.

Jencks is concerned with the USA and with the distribution of status and income. He therefore does not use the framework I have suggested for putting transmitted deprivation into the perspective of varying societal levels of affluence and rates of mobility. He also has nothing new to tell us about the 'index of association' approach to defining and analysing a transmitted deprivation cycle. This is so, in part, because he was using the type of data which do not permit sufficient isolation of the category of deprived persons. It is characteristic of the samples used in this kind of demographic analysis (for example, Blau and Duncan, or Glass, or the present Oxford mobility study) that they, being concerned primarily with mobility and not deprivation as such, have both too few numbers and insufficient data on the relevant subsample of the population. The general approach certainly permits in principle a testing of the cycle hypothesis in the way that I have interpreted it. Nevertheless, it would be necessary to use a research design which included an adequate sample of the very poor, was not confined to males, and asked more questions about genetic and social transmission as well as about sources and amount of income.[7]

In any case, the 'index of association' approach is only a first step. The theory of a deprivation cycle or a 'culture of poverty' also includes hypotheses about the nature of transmission. To analyse these processes other methods are necessary. One of these is path analysis, of which Jencks gives an excellent exposition in his Appendix B (pp.320 ff.). Duncan (in Moynihan 1968) has used the same method for a penetrating analysis of the educational attainments, occupational status and income of Negroes in the United States. A similar model could be used in further explorations of transmitted deprivation. Again, Jencks is not directly concerned with this topic but he does have conclusions and interpretations as to what factors are important in determining the occupational and income position of individuals.

At this point, however, a serious criticism of Jencks could be mentioned which has been brought by James Coleman and others.[8] As Coleman puts it:

> The argument that equalizing education will not result in equalizing incomes is one about the relation of variance of education to variance in incomes, at a societal level. Yet all the analysis is concerned with relations at the individual level. Whether education is related to income at the individual level is quite independent of whether variance in education is related to variance in income at the societal level.

Table 4 Significant differences between indices of association for total male sample, Britain 1949

| Father's status category | 1 | 2 | 3 | 4 | 5 | 6 | 7 |
|---|---|---|---|---|---|---|---|
| 1 | 13·158 | | | | | | |
| 2 | * | 5·865 | | | | | |
| 3 | * | * | 1·997 | | | | |
| 4 | * | * | ‡ | 1·618 | | | |
| 5 | * | * | * | * | 1·157 | | |
| 6 | * | * | ‡ | ‡ | * | 1·841 | |
| 7 | * | * | ‡ | † | * | ‡ | 2·259 |

Notes: * Significant at 1 per cent level
† Significant at 5 per cent level
‡ Not significant
Source: Glass (1954).

The same point could be made about all factors which determine life chances.

This point is relevant because, as I understand it, the theory of transmitted deprivation is usually phrased in individual terms, whereas the wider problem of poverty has to be explained in terms of the structure of society. As Duncan says in *On Understanding Poverty* (Moynihan 1968), 'if there were any chance that the slogan-makers and the policy-builders were to heed the implications of social research, the first lesson for them to learn would be that *poverty is not a trait but a condition*'. There is always the danger of the fallacy of composition in analyses which turn situational factors into individual attributes. But this is in effect what Jencks does in his book. Thomas Pettigrew has also made the same criticism.[9] 'While inequality among individuals is given intense attention, group inequality is virtually ignored. When race and class are considered, they are typically treated as characteristics rather than as group phenomena around which inequalities in a complex, heterogeneous industrial society is best judged.'

Nevertheless, Jencks does have some conclusions about the processes involved in the life cycle which need to be considered. First, he concludes (p. 159) that the most important determinant of educational attainment is family background. Obviously, this conclusion has also been well documented for Britain. Second, he finds (p. 256) that 'the character of a school's output depends largely on a single input, namely, the characteristics of the entering children. Everything else – the school budget, its policies, the characteristics of the teachers – is either secondary or completely irrelevant.' Most observers would want to qualify this finding, at least in application to Britain. Third, Jencks notes that educational opportunities are unequally distributed. There is inequality of access, to types of schools and types of curriculum, and there is inequality of resource. Clearly, these findings are also applicable to Britain.

Next, Jencks considers the distribution of cognitive ability or test scores. He calculates, in his brilliant Appendix A, more convincing figures and offers a more interactionist explanation of test scores than is to be found in the work of the hereditarians (Jensen, Eysenck, etc.). He then goes on to draw conclusions about the effect of schooling. Inequality in test scores might decline, he asserts, by 9 to 19 per cent if the amount of schooling and the quality of schooling were equalized and by 6 per cent if everyone's economic status were equalized. He says (p. 109) that 'additional school expenditures are unlikely to increase achievement and redistributing resources will not reduce test score inequality.'

Fifth, addressing himself to the American desegregation problem, he calculates that eliminating racial and socioeconomic segregation in schools might reduce the test score gap between the black and white children and between rich and poor children by 10 to 20 per cent (p. 109).

He then goes on to the connection between education and occupation and finds that occupational status is strongly related to educational attainment, though there are 'enormous' differences among people with the same amount of formal schooling. But he notes that the fact that blacks hold poorer jobs is largely due to direct racial discrimination.

Then he sums up his main thesis (p. 226) with the generalization that 'neither family background, cognitive skill, educational attainment, nor occupational status explains much of the variation in men's incomes'.

Now if these findings, and especially the last summary conclusion, are taken as they stand, then the implications for the theory of transmitted deprivation are strange. If there were such a cycle, presumably very little could be done about it by trying to change either the family background or the intelligence or educational qualifications or jobs of its victims. All that could be done with any practical efficacy would be to raise the income of those found in the cycle. Yet I do not imagine that many social scientists or policy-makers will find this conclusion easy to accept (see James Meade, p. 172 below, for a further discussion of this point). They will therefore not be surprised to find that Jencks's work has been increasingly criticized by subsequent reviewers.

Within the context of commitment to a more equal America, Jencks's main aim was to attack the orthodoxies of the Kennedy/Johnson Washington of the 1960s which included the doctrine that poverty was a vicious cycle of family failure to equip children with the skills needed to rise into and keep good jobs, but more especially that educational reform was the best mechanism for breaking the intergenerational curse. It was the educational link in the cycle to which he and his colleagues gave most attention.

The book is very largely, despite its title and its concluding chapter, concerned with the determinants of educational success. Hence the major finding that schooling explains only about 12 per cent of the variance in income. Quite apart from the criticism that the explanation for the inequality of income accruing to jobs is not what would explain who happens to occupy them, there is also considerable doubt as to whether the path analysis technique on which Jencks relies has not resulted in an underestimation of the effect of schooling on both occupational status and income distribution. And, even more serious, this form of analysis leaves a large unexplained variance in income which, in various formulations, he attributes to luck and a capriciously, if not arbitrarily, distributed competence. Jencks has since admitted that this was a serious error. There is no reason to suppose that because we cannot identify the causes of a result we should infer that the causes are random rather than part of some systematic feature of inheritance or society. On the other hand, that systematic feature is by no means necessarily a cycle of transmitted deprivation. It has to be decided by empirical inquiry whether 'luck' and the other processes involved with the unexplained variance may co-vary with education and make schooling more important as a determinant of life chances than Jencks's book indicates.

I am not alone in believing that Jencks has overestimated the role of chance and underestimated the effects of family background, ability and education on adult attainments. Apart from James Coleman and Thomas Pettigrew, whom I have mentioned, there is similar criticism to be found in the review by William Sewell.[10] Sewell writes:

> As should be clearly understood, much of the analysis upon which Jencks's conclusions are based comes from the use of linear regression techniques. Although his use of these techniques is straightforward, the standard he employs for assessing the importance of an independent variable or of several variables arranged in a causal sequence, is always the percentage of variance explained in the dependent variable ($R^2$). Thus, little attention is given to the role of a variable or series of variables in explaining the complex process by which achievements take place over the course of the life cycle of individuals. Actually, the scientific importance of a variable may reside more in its

interpretative role in a causal process than on the amount of variance it explains. Thus, the value of the Blau–Duncan models and the extensions of them by me and my associates inheres more in their ability to elucidate the achievement process than in the fact that they explain from 25 per cent to 60 per cent of the variance in educational and occupational attainments.

This overconcern with $R^2$ (and increments in $R^2$) accounts also for Jencks's easy rejection of many relationships that by usual standards in quantitative social science would be considered quite important. He really comes down to setting a standard that says a causal variable (or set of causal variables) is unimportant if it does not explain most of the variance in the dependent variable of interest. Aside from being an unrealistic standard for the empirical world in which social sciences operate, Jencks equates residual variance with luck, which leads him to the conclusion that luck is more important in determining men's fate than their social origins, their cognitive skills, and their educations.

Although I am willing to credit luck with an important role in achievement, I must point out that the amount of residual variance in any regression model may be due to at least several other sources: (1) unreliability in the measurement of the independent variables, (2) failure to include in the regression model other exogenous and intervening variables that would make a significant contribution to variance in the dependent variable, and (3) failure to adequately define and measure the dependent variable. . . .

It can easily be demonstrated that a large increase in the variance explained in educational and early occupational attainments can be gained by adding a small number of social psychological variables to the basic Blau–Duncan model but, more important, these variables elucidate the achievement process by showing how socioeconomic background is mediated by these variables. (See Sewell *et al.* 1969, Sewell and Hauser 1972.)

Regression analysis and path analysis, important as they are for discovering and elucidating the extent and nature of the hypothesized cycle of transmitted deprivation, can never be adequate substitutes for true experiments. Jencks's lack of caution on these points produced the premature conclusion that the effects of family background, cognitive skills and schooling are negligible in determining later achievement. In one way Jencks's conclusions would lead us to dismiss the theory of a transmitted deprivation cycle. But even if we rid ourselves of the Jencksian error it would still be necessary to devise a clear experimental design for testing the theory. This would require long-term experiments in which, ideally, children of known genetic and social origin would have to be assigned to schools at random, or, as a substitute for direct experimentation, there would have to be careful measurement of this and other interfering variables so that their effects could be controlled statistically.

A final criticism to be noted is one which emanates from those concerned with the inequalities of ethnic and class minorities. Because Jencks concentrates his attention on inequality of opportunity rather than on inequality of result and on individuals rather than on groups his book has tended to underemphasize the importance of their efforts to improve the relative opportunities for groups which can be materially altered by educational reform even though it is also true that individual inequality would remain and that not all group inequality is caused by family and educational inequality.

However, although I have emphasized the shortcomings of Jencks's book from the point of view of the study of transmitted deprivation, it is also appropriate to praise it. The book has enlivened a central social debate of our time. Jencks has effectively demonstrated that some recently fashionable theories of educational aids to equality are dubious. He probably underemphasized the possibilities of greater fairness between social groups through educational reform. And he exaggerates the degree to which people had believed in education as an effective mechanism for the creation of equality. He has, however, cleared the way to a more constructive debate as to how much inequality there is, what reduction in it is possible and through what mechanisms. Among these it may well be that transmitted deprivation plays a relatively minor part in reproducing inequality in each new generation.

## Notes

1  The excerpts chosen come from a rapidly increasing literature. Many other examples were excluded by the exigencies of practical requirements for a reasonably short and commercially viable book.

2  Husén's book contains an excellent discussion of genetic and environmentalist approaches to the distribution of ability between groups.

3 Part II covers population control and fertility. Part III covers intelligence and social stratification. Part IV covers IQ, genetics and race.

4 See the article by Oliver Gillie in the *Sunday Times* (24 October 1976) and subsequent correspondence in *The Times*. With respect to the high probability that Burt invented his collaborators Miss Conway and Miss Howard, I can add a footnote. Burt wrote to me in 1959 asking me to reply to a criticism (of the article reprinted below in Chapter 11) to be published by 'Miss Conway' in the *British Journal of Statistical Psychology* which he edited. I did so. Burt then wrote another article endorsing 'Miss Conway's' view and gave me no opportunity to reply

5 I rely in the following pages on a memorandum which I was invited to submit to the Social Science Research Council. My thanks are due to the Council for permission to use the material here.

6 Blau and Duncan have seventeen socioeconomic levels based on median income and education. The sixteenth and seventeenth levels are farmers and farm labourers. The fifteenth category (i.e. the lowest of the non-farm levels) of 'other labourers' had a self-recruitment percentage of 9·9 – i.e. 9·9 per cent of the sons of these labourers themselves became labourers. The details are shown in Table 2.

7 Incidentally, the practical task of securing a high response rate from the highly deprived elements in the population is severe.

8 See *Amer. J. Sociol.*, **78**, 6 (1973). See also the symposium of reviews of Jencks in *Harvard Educ. Rev.*, **43**, 1 (1973).

9 In *Amer. J. Sociol.* (see note 8 above), p. 1528.

10 Ibid., p. 1532.

# PART ONE

*Approaches*

# 1 Geneticism and environmentalism

*J. M. Thoday*

(From J. E. Meade and A. S. Parkes (eds), *Biological Aspects of Social Problems*. London, Oliver & Boyd, 1965. pp. 92–106.)

In 1894, in one of the classics of genetics, Bateson wrote the following:

It is especially strange that while few take much heed of the modes of Variation of the visible facts of Descent, everyone is interested in the *causes* of variation and the nature of 'Heredity', a subject of extreme and peculiar difficulty. In the absence of special knowledge these things are discussed with enthusiasm even by the public at large. (Bateson, 1894.)

Bateson was then arguing for less talk, opinion and prejudice and some study, experimental study, of variation. He practised what he preached and was rewarded by becoming one of the most distinguished of the early geneticists: it was, in fact, he who gave the science its name.

Since then our knowledge of genetics has advanced considerably. Nevertheless, in many respects and in many fields, what Bateson complained of is still true, most notably in respect to psychological characters in man, where the advance of genetic knowledge has been slight.

Bateson himself, like all the early geneticists, was of course reared in the traditional attitude that gave exaggerated importance to adult form, and fell himself to some extent into the trap it creates. He and many others thought in terms of genes for particular adult colours or forms, and, as Stebbins (1963, p. 312) has pointed out, 'started the kind of thinking in which some scientists who are not geneticists indulge, when they speak of genes for an eye, a nose, or for musical genius or the ability to remember mathematical formulae.'

It is this kind of thinking that leads to discussion in terms of heredity *or* environment, and to the extreme environmentalism of many physiologists, psychologists and educationalists, and, partly by reaction, to the antithetic extreme geneticism of others.

An illustration of this deficient understanding may be taken from the Newsom (1963) report. I would first say that this is one of the most imaginative of the reports on educational problems I have seen. It contains, in Chapter 2, an unexceptionable statement: 'There were well over two and three-quarter million boys and girls in maintained secondary schools in 1962, all of them individuals, all *different,* (italics mine). This statement sets the tone of the report, which frankly recognizes that people differ, and that different people may *need* different treatments.

This is excellent. And it is in no sense of criticism of the Newsom Committee that I draw attention to my next quotation, which illustrates how little the role of genetic variation is really understood. I quote:

> The results of such investigations increasingly indicate that the kind of intelligence that is measured by the tests so far applied is largely an acquired characteristic. This is not to deny the existence of a basic genetic endowment; but whereas that endowment, so far, has proved impossible to isolate, other factors can be identified. Particularly significant among them are the influences of social and physical environment; and since these are susceptible to modification, they may well prove educationally more important.

Now there is no statement of fact to disagree with there. And no one would disagree that the modifiable factors are more important in the practical sense. What I want to draw attention to, however, are two statements whose meaning will not pass analysis. The first statement is that the intelligence measured by these tests is largely an acquired characteristic and the second is that the basic genetic endowment has proved impossible to isolate.

About the first let us be clear right away. No characteristic is largely acquired. Every characteristic (apart from the breeding potential of an individual[1]) is entirely acquired. Every character of an individual is acquired during the development of that individual. Likewise every character is genetic, for to acquire a character during development in any particular environment the individual must have the necessary genetic endowment; and this is as true of learned characters as of those that are not apparently

learned. But knowledge of the genetic endowment is of little
without knowledge of the environmental circumstances
Every character is both genetic and environmental in origin. Let
us be quite clear about this. Genotype determines the potentialities
of an organism. Environment determines which or how much of
those potentialities shall be realized during development. The
doctrine of fixed abilities is nonsense.

The second statement is more perplexing. The assertion that
'the basic genetic endowment has proved impossible to isolate'
immediately gives rise to the idea that it may in principle be
possible to isolate genetic endowments, as if we could pick out the
gene or a group of genes for an IQ of 126·5. The statement can
only have meaning in the context of genes for IQ or genes for
musical genius, and given such a concept of genetics we are
straight in the situation Stebbins complained of. Furthermore, it
provides a context in which proof that environment can affect a
character is easily taken as proof that genotype does not.

Let us then consider what it is that genetics has to say about
such characters as height or IQ in man, yield in corn, or hair
number on flies, in order to approach some feeling for the
complementary roles of genotype and environment in determination of the developmental process concerned.

First we must remember that all such characters are relative.
When we say a man is 5 feet 6 inches high we are saying that he is
6 inches shorter than two standard yards, tucked away in the
National Physical Laboratory or the British Museum or wherever
the standard yard is kept. Another man is 6 feet 3 inches, and these
measures have only two kinds of use. One is to determine which
sized door, trousers or what you will will be relatively more
suitable for one or other man. The other is to enable us to say one
man is taller than the other. The taller man may be said to have
the character tallness or to be tall, but this is purely a shorthand
statement meaning he is taller than some or taller than average.
Gulliver was tall in Lilliput but dwarf in Brobdingnag. His
tallness or dwarfness had no more to do with him than with the
other members of the society in which he found himself. The
character is an abstraction and we give it meaning more than as a
mere relationship at our peril. Talk of genes for tallness, or of
isolating a basic genetic endowment for tallness is dangerously
misleading.

It is the more so with intelligence, musical ability, mathematical ability. When we try to measure these we are measuring relations, and we say nothing about the individual as such except to place him on a scale relative to other individuals. We can say nothing useful about intelligence as measured, except in relation to a population.

When considering any such character, therefore, we are considering a population. When asking about the genetic factors or the environmental factors that influence intelligence, or any other character, we are asking about the causes of variety in a population. And whatever we say of an individual can only be a statement concerning the differences between that individual and others in the population, whether what we say concerns his intelligence as measured, or the causes of his relatively high or low IQ.

Let us then consider a character about which much is known, hair number in flies, which can provide a model to show how characters of this kind may be influenced. Let us remember we can only talk about differences of hair number. Let us ask what factors may affect hair number.

First, of course, any population we may consider has a mean or average hair number.

Second, it shows variety of hair number, it has variance.

Third, the variance has causes, many causes. (The concept of *a* cause of *a* character has no place in genetics and should have none in biology.)

Fourth, different populations may differ in mean, variance or both.

Fifth, if two populations differ in mean, there are causes of the difference.

Sixth, if two populations differ in variance, then there are more, or more effective, causes of variety in that which has the greater variance.

Seventh, unless the populations have been artificially produced by special breeding programmes designed to eliminate genetic variety, the causes of variance are *always* both genetic and environmental, and the variance can be partitioned into three components, genetic variance, environmental variance and variance arising from genotype-environment interaction. In other words, we may classify the causes of variance into these three groups and assign them relative importance. This is true of every continuously varying character that has been adequately studied of any species of outbreeding organism. It is undoubtedly true of IQ variation.

What may differ from population to population, from character to character, or from time to time, is the relative importance of these three classes of cause, or, more precisely, the proportion of the variance that is assignable to each of these three causes. The proportion of the variance assignable to genetic causes, which we call the heritability of the chatacter, may vary for either of two reasons. First, it will increase with increase of, and decrease with decrease of the amount of genetic variety in the population. Second, it will decrease with increase of and increase with decrease of the variety of environments in which the individual members of the population develop.

This is of the utmost importance. The more uniform the environment, the greater will be the importance of genetic variety. In other words, the greater the equality of opportunity the more important will differences of genetic endowment become.

This is all rather simple. But relations become more complex when we consider the third factor – genotype-environment interaction, a factor whose importance will increase with the complexity and variety of the genetic and the environmental causes of variation, and which must therefore be of maximum importance in characteristics that vary with social background, home background, etc.

Here I want to give an example from experiment, and also to illustrate the concepts with two simple models.

The experiment concerns hair number of flies. A population of flies was classified for hair number, and two samples, exactly comparable, were taken from it. Each sample was used to generate a new generation, by mating together the least hairy female and male, the next least hairy pair and so on to the most hairy pair. The average hair number was obtained from the progeny of each pair. The two samples' progenies were grown in different environments, one at 20° C, the other 25° C, the parents having all been grown at 25° C. The results are summed up in Figure 1.1, in which the statistical relation between parent hair number and progeny mean hair number is shown for each environment.

Two things are immediately obvious. First, the lower temperature produces higher hair numbers. The environmental difference alters relevant developmental processes effectively. Second, the higher hair number parents produce higher hair number progenies.

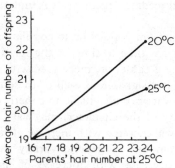

*Figure 1.1* The relation of offspring hair number at 20° and 25° C to parent hair number at 25° C. The differences of regression coefficient and mean both differed significantly between environments.

Relevant developmental processes are effectively altered by genetic variety in the samples. Two less obvious things are important. In neither environment is hair number completely inherited, that is to say that the extreme hair number parents produce offspring with less extreme average hair numbers. Second, the relative effect of genetic difference varies with environment, and the relative effect of the environmental difference varies with genotype. In fact the temperature of development made no difference at all to the progeny of the low hair number parents, but a good deal of difference to the progeny of high hair number parents. This is an example of genotype-environment interaction. It leads to different heritabilities in the differing environments. When the flies developed at 20° C, 40 per cent of the variance was genetic, the remaining 60 per cent being largely assignable to unknown environmental variables other than temperature. When the flies developed at 25° C, only 20 per cent of the variance was genetic, 80 per cent being assignable to residual environmental factors.

I give this example not only to show what genotype-environment interaction means, but also to underline the point that heritability – that is, the proportion of variance that is genetic – is not a quantity that belongs to a character in a species, but to a population in its environment and will vary according to the population and the environment. Controversies concerning the heritability of IQ in man, arising from different estimates obtained by different workers from different samples, are rather silly. For example, it is to be expected that heritabilities estimated from study of twins reared together should differ from those derived from study of twins reared apart.

To return to genotype-environment interaction, the situation can be more striking than that illustrated above, for a change of environment can actually change the order of genotypes.

For a simple example I turn to yield of wheat, two varieties being compared over two years – that is, in two environments. In one year Turkey Red yielded significantly less than Big Frame. In the next year significantly more. Neither environment is best, neither variety is best. One variety is better in one year, the other in another year.

Two imaginary model situations will illustrate the significance of these considerations. Consider IQ, and imagine a population genetically simple enough to contain only five kinds of genotype each equally frequent. Set it in a situation so simple that there are only five kinds of environment, all likewise equally frequent.

Without genotype-environment interaction we might get phenotypic IQ classes as shown in Table 1.2.

Table 1.1 *Relative yields of two varieties of wheat in two years*

| Year | Variety | |
|------|---------|---|
|      | Turkey Red | Big Frame |
| 1913 | 100 | 107 |
| 1914 | 100 | 85 |

*Source:* Hayes and Immer 1942, Table 38, p. 299.

Table 1.2

| Environment | Genotypes | | | | |
|-------------|-----------|---|---|---|---|
|             | A | B | C | D | E |
| A | 80 | 85 | 90 | 95 | 100 |
| B | 85 | 90 | 95 | 100 | 105 |
| C | 90 | 95 | 100 | 105 | 110 |
| D | 95 | 100 | 105 | 110 | 115 |
| E | 100 | 105 | 110 | 115 | 120 |

Table 1.3

| Environment | Genotypes | | | | |
|---|---|---|---|---|---|
| | A | B | C | D | E |
| A | 80 | 90 | 100 | 110 | 120 |
| B | 120 | 110 | 80 | 90 | 100 |
| C | 90 | 80 | 120 | 100 | 110 |
| D | 110 | 100 | 90 | 120 | 80 |
| E | 100 | 120 | 110 | 80 | 90 |

In this simple situation in which there is no genotype-environment interaction it is obvious that, if we wished to maximize IQ, we should give everyone environment E. If on the other hand our motives were egalitarian, we should give the A genotypes environment E, and the E genotypes environment A. Or if for economic or other reasons we were only able to give environment E (the best environment from this point of view) to a few, and wished to have as many high IQs as possible in the population, we should have to devise selection techniques so that we could choose E genotypes for the E environment.

By contrast, with genotype-environment interaction we might have something like Table 1.3.

Now, because I have chosen an extreme situation in which all the variance is interaction variance, we have no environment better than any other, and no genotype better than any other. Each genotype requires a different environment if it is to develop the highest IQ it is capable of. A requires environment B, B requires E, C C, D D and E A.

Now no natural situation is as extreme as these, and no natural population as simple. In natural situations there will be some interacting and some non-interacting variations of both genotype and environment, so that we must think of a model somewhere between the two I have given. But in addition we must remember that natural populations are of fantastic genetic complexity and natural environments are also complex, especially the natural environment of civilized man, with all the variation in social, economic, educational, home and school background, etc.

It is becoming more and more clear, the more experiments we do on the genetics of natural populations, that the old idea, itself relating to the biological type concept, that populations are genetically rather uniform, there being by and large one normal or wild type, along with many abnormalities, each rare, is totally misleading. Normal flies, or normal men, comprise an extensive array of differing genotypes, genotypes that differ in their effects on any characteristic we like to study closely enough. So extensive is this variety that we may say without exaggeration that, apart from identical twins, no two individuals are, or ever have been genetically exactly alike. Each human being is genetically unique. We are gradually coming more and more to understand why this should be so; but the fact and its immense implications, still less the value of such variety to society, are barely recognized. Superimposed on this genetic uniqueness, we have the uniqueness of environment of each individual during development, some of the environmental variation itself having genetic variation as part cause. Unique genotype and unique environment are interacting in the development of each individual in unique ways, and, though we must classify individuals into groups for scientific, administrative or educational purposes, we ignore this uniqueness to our great loss and at our peril, and it makes nonsense of segregation of races justified solely on the basis of differences of average even if the average differences may be real.

Consider a simple example of this variety, using a clear case of a genetic difference in the development of which environmental variance plays little part: red–green colour-blindness. First, this is a genetic difference leading to a difference in perception of the outside world. A colour-blind individual meets a different environment, experiences a different world, from a colour-perceptive individual. No doubt there is loss in this for him, though it must be loss he cannot appreciate any more than we can. But there is also gain. One factor contributing to the success of *Drosophila* experiments in my department is the colour-blindness of one of our senior workers. He can classify some phenotypes important to our breeding programmes which the rest of us cannot. We can classify some that he cannot. He tells me also that colour-blindness has advantages in another field, the dyeing industry. For certain jobs which involve careful matching of tints of the same colour, colour-blind individuals are in great demand. Both genotypes have value, and society is the better for the existence of this variety.

We may even question whether a complex, integrated society such as ours is possible without genetic variety, for the integration of a society clearly requires differences as well as similarities between the individuals that have to be integrated. The rigid social systems of ants and bees depend largely on age structure and rearing differences and rather little on genetic differentiation apart from that distinguishing the sexes of bees. But it is striking that among primates even so relatively simple a social organization as that of baboons seems to depend on uniqueness of the individual. All members of the group are individually recognizable by each other. It seems doubtful whether the complexity of social organization upon which human success depends could have developed without the extensive genetic variety human populations contain.

Another example I wish to consider illustrates an even greater complexity of genotype-environment interaction: this time from work on mice. De Fries (1964) has recently reported the following facts. Pregnant female mice of various inbred strains were subjected to stress conditions in swim tanks, sound chambers, etc. Studies were made of behaviour of their offspring and those of controls. Significant effects of treatment of mother on behaviour of offspring were observed, but these effects of treatment varied both with genotype of mother *and* with genotype of offspring. Here we see a beautiful example of mother's genotype interacting with mother's environment to affect offspring's environment, and offspring's environment interacting with offspring's genotype to determine the mode of offspring's development and hence the observed characteristics of offspring. Such situations, made more complex by considering interaction between parent and offspring in the house, parent, offspring and teacher in the school, and so on, provide the sort of model we must think about when considering biological variation in relation to educational policy.

An old cardinal principle of education takes these complexities into account. It is the principle that each pupil requires individual attention, in other words a unique environment. Yet at the present time we see more and more stress on a different principle, that it is unfair if some get better opportunities than others. This would be fine if it were not coupled with an implicit assumption that the best opportunity for one would be the best for all. Thus arises much of the disturbance about selection for education, and the unfairness of giving some 'better' opportunities in grammar schools and universities, controversy about the eleven-plus, and so on.

All these attitudes are tenable only if we assume that there is such a thing as human nature. But there is not. There are as many human natures as men. To do his best for society, or to get the most from life, or to have the best compromise between the two, each individual would require a unique, carefully chosen environment. This we cannot provide, if only because we do not know enough. But in one sense at least selection techniques for educational streams are an attempt to do the best we can. Some injustice arises from selection because our selection techniques are of limited efficiency, but we know, both on the theoretical grounds, and from direct experience, that injustice can also arise through lack of selection, or from giving individuals a so-called better environment which is unsuitable for them.

Let me end by pleading that geneticism and environmentalism both may die a rapid death; that we begin to realize that the question 'Is this character inherited or is it acquired?' can never have an answer other than 'No', because the question has no meaning; and that we take into account, in social and educational theory, policy and polemic, the biological fact of the uniqueness of the individual, implying as it does that the needs of each individual are unique and that it is *just* to treat different people differently so long as each is treated as well as possible.

Nobody has the equipment necessary for discussion of these topics unless he thoroughly understands the concept of genotype-environment interaction in development, and appreciates that all men are different, different from conception.

## Notes
1 i.e. the potentialities that an individual can hand on to his offspring.

# 2 Natural selection in mankind

## T. Dobzhansky

(From G. A. Harrison and A. J. Boyce (eds.), *The Structure of Human Populations.* Oxford, Clarendon Press, 1972. pp. 213–33. Reprinted by permission of the publisher.)

### Fitness

Darwin's beautifully simple and clear definition of natural selection was 'preservation of favourable variations and the rejection of injurious variations' (1859).

Natural selection is a sequel to the struggle for existence. This struggle includes 'not only the life of the individual but success in leaving progeny'. The stress is nevertheless on mortality. Darwin acknowledges the doctrine of Malthus as the source of his own: 'There is no exception to the rule that every organic being naturally increases at so high a rate that, if not destroyed, the earth would soon be covered by the progeny of a single pair.' We must 'keep steadily in mind that each organic being is striving to increase at a geometric ratio; that each at some period of its life, during some season of the year, during each generation or at intervals has to struggle for life and to suffer great destruction'. If this makes one feel squeamish, 'we may console ourselves with the full belief that the war of nature is not incessant, that no fear is felt, that death is generally prompt, and that the vigorous, the healthy, and the happy survive and multiply'.

The emergence of genetics has shifted the attention of evolutionists to the transmission of genes from one generation to the next. Selection occurs when the carriers of some gene variants leave more or fewer surviving descendants relative to the carriers of other variants. Or, to put it differently, natural selection is differential perpetuation of genetic variants from generation to generation. The carriers of a certain variant may be more viable, or more fecund, or sexually more active, or reach sexual maturity earlier, or have a longer reproductive period than do the carriers of another variant. Any one, or any combination, of these advantages may influence the contribution of the carriers of a given genotype in one generation to the gene pool of the next generation. The magnitude of the contribution of the carriers of a given genotype relative to those of other genotypes in the same population is the Darwinian fitness of the genotype. Adaptive value or selective value are alternative terms used synonymously with Darwinian fitness. Natural selection takes place when a population contains two or more genotypes with different Darwinian fitnesses, in the environments in which this population lives.

Darwinian fitness should not be confused with fitness in everyday language. This is why 'Darwinian' is added. Darwinian fitness is reproductive fitness. In principle, natural selection could take place without differential mortality. Imagine a population in which all the children born survive to maturity. However, some prospective parents may remain celibate, others produce few children, and still others many children. If the numbers of the children born are correlated with the genetic constitutions of the parents, differences in Darwinian fitness are present, and selection is taking place. Or else, suppose that everybody is married and has the same number of children, all of whom survive. Selection could still operate, if some people would have children at a younger and others at a later age, and if the age of childbearing is a function of the genetic constitution.

Of course, selection can also operate with differential survival of the children born, regardless of whether genetic variations in fecundity or in the childbearing age are present or not. Crow has proposed a very simple measure of 'the opportunity for selection'. This is $I = V/W^2$, where $V$ is the variance of the numbers of children born per couple, and $W$ is the mean number of children. The greater is $I$ the more opportunity there is for selection to operate. The value, $I$ has two components, $I_m$ due to differential mortality, and $I_f$, due to differential fertility. As expected, the values of $I$ are diverse in different human populations. What is, however, interesting and rather surprising is that, by and large, the $I$s are greater in populations with small average numbers of children per couple than in those with many children. This is so because the variances of the numbers of children per family grow relatively less rapidly than do the squares of the average numbers of children. In other words, large families are by no means necessary for natural selection to occur. Variations in the numbers of children per family are sufficient, provided, of course, that they are genetically conditioned (Spuhler 1963).

It must be stressed that Darwinian fitness is a relative measure. A genotype may have a higher or a lower Darwinian fitness than other genotypes. For example, colour-blind or myopic individuals were probably at some disadvantage compared to those with 'normal' vision in tribes of hunters and food gatherers. Diabetics are at a disadvantage in modern well-fed societies, notwithstanding the ministrations of modern medicine. It is certain that colour-blindness and some forms of myopia and diabetes are genetically conditioned. Since these defects reduce the Darwinian fitness of their carriers in certain environments, natural selection did, and probably still does, discriminate against them, and favours what we call 'normal' eyesight and carbohydrate metabolism.

Imagine, however, that in mankind everybody is myopic, or colour-blind, or diabetic. It is at least conceivable that a human species consisting entirely of myopic, or colour-blind, or diabetic individuals could have survived. It would then be meaningless to say that these characteristics reduce the Darwinian fitness of their possessors, because we would have nothing to compare them with. Nobody would even suspect that he is a carrier of a genetic defect. If a person with what we call normal vision appeared by mutation in a colour-blind mankind, he would be regarded a possessor of an extraordinary ability or possibly a sorcerer. If his progeny inherited his vision, their Darwinian fitness might be higher than that of most of their conspecifics, and natural selection would increase their incidence from generation to generation.

## Adaptedness

Every living species or population is adapted to live in certain environments. This statement is almost tautological: a species without adaptedness would be extinct. But it is not quite a tautology, because there are different degrees of adaptedness. For example, California condor and whooping crane have very narrow ranges of adaptedness, as manifested by their being reduced to small numbers of individuals. By contrast, the Norway rat and house sparrow are flourishing in a variety of climates and on a variety of foods. The human species has an unequalled adaptedness, owing to the ability to control the environments and to devise new environments by means of culture and technology. Since the ability to develop and to maintain culture has a genetic basis, the adaptedness must be considered a product ultimately of the human genetic constitution.

It seems intuitively obvious that the adaptedness of modern mankind is superior to that of mankind of palaeolithic times, or of the australopithecine ancestors. One can also surmise that uncontrolled population growth and despoilation of the world environments may seriously lower man's adaptedness. The difficulty with the concept of adaptedness is that there are thus far no good methods for its quantification. However difficult it is in practice, the measurement of the Darwinian fitness of human genotypes is theoretically straightforward. Data must be obtained on the reproductive performance (including, of course, the survival of the progeny) of the carriers of the genetic variants under study, in relation to the performance of other genotypes at the same time and in the same environment. Thus, the Darwinian fitness of achondroplastic dwarfs in Denmark is roughly 0·1 of that of the non-dwarfs (Mørch 1941, Popham 1953). This value may, of course, change with time, and it may be different in different social and physical environments.

As to the adaptedness of a population or a species, the numbers of individuals and the biomass are clearly relevant. On these criteria, California condor and whooping crane are far below the house sparrow and the rat, and these are below man. But this is not the whole story. A statistic called the malthusian parameter, or innate capacity for increase, has been proposed to measure the rate of the population growth in an environment in which the living necessities are not limiting (Andrewartha and Birch 1954). This statistic, denoted $r_m$, tells us, in essence, how rapidly the number of individuals of a species can increase in a certain environment before its carrying capacity becomes insufficient.

The innate capacity for increase is vastly greater in many lower organisms, such as insects and micro-organisms, than in man. It can, however, be argued that for such very different organisms, living in quite different environments, comparison of the $r_m$ values is really not informative. It is more meaningful for populations of the same species or closely related species. Thus, the rate of growth and the doubling time of human populations has speeded up in recent centuries, particularly since the Industrial Revolution, and it is still becoming more rapid.

Furthermore, not all human populations and races have been growing equally rapidly. Some have, in fact, declined. It is reasonable to say that the adaptedness of the human species, and particularly of the populations living in technologically advanced environments, has increased. This is not contradicted by the fact that the runaway population growth will be a calamity – any environment is bound eventually to become limiting for any species, no matter how excellent its adaptedness.

**Adaptation and adaptability**

Adaptation is a process whereby the state of adaptedness is achieved. 'Adaptation' is used also as an abbreviation for adaptive trait, i.e. for a structure or a function which contributes to the adaptedness. This double meaning is unfortunate, but it is probably too firmly rooted in the biological vocabulary to be rectified easily. Adaptation as a process may be physiological or genetic. An individual adapts physiologically to low oxygen pressure by some changes in the composition of his blood, and to sunlight exposure by deposition of more skin pigment. Genetic adaptation of a plant population to the presence of parasitic fungi, or the presence of salts of heavy metals in the soil, takes place by selection favouring the genotypes which confer immunity to these noxious agents. Similar selective processes have presumably occurred in human evolution in response to pathogens of various sorts.

Natural selection results in adaptation when genetic variants which confer superior adaptedness have a Darwinian fitness higher than do variants of lesser adaptedness. As a rule, superior Darwinian fitness goes together with superior adaptedness, and vice versa. This statement is by no means trivial. Exceptions are known when greater Darwinian fitness turns out to be the property of genotypes which give rise to maladaptive traits. A striking example are some of the alleles at the complex $t$ locus in mice, extensively studied by Dunn and his collaborators (Dunn 1964 and other works). These $t$ alleles are lethal to homozygotes, and yet males heterozygous for $t$ produce many more sperms carrying $t$ than its normal allele. This makes $t$ have a superior Darwinian fitness, and yet a population with high frequencies of $t$ suffers heightened embryonic mortality. Another example is the 'sex ratio' conditions, carried in many $X$ chromosomes in natural populations of several species of *Drosophila*. A male with such an $X$

chromosome transmits it to his entire progeny, instead of to half the progeny as in normal males. If not counteracted by some other factor, the 'sex ratio' condition would spread in the population, until the latter becomes unisexual, which in an organism incapable of parthenogenesis means extinction. Though not known with certainty, genetic variants like the *t* alleles in mice or the 'sex ratio' in *Drosophila* may exist also in man.

Neither the Darwinian fitness nor the adaptedness are immutable attributes of a genotype. They depend on the environments, and consequently may shift upwards or downwards at different times and in different places. By and large, the Darwinian fitness and the adaptedness are positively correlated; if they were uncorrelated, life on earth would probably have become extinct long ago. Yet even apart from the rather exceptional situations mentioned above there is a fundamental constraint on the ability of natural selection to conserve or advance the adaptedness of a population or of a species. This is simply that natural selection has no foresight. It selects what is advantageous, and has a high Darwinian fitness where and when the selection operates. In short, natural selection is opportunistic.

While it brings a short-term advantage, opportunism may be disadvantageous in the long run. It entails the risks of failure of adaptedness and of eventual extinction. After all, extinction is the commonest destiny of most evolutionary lines, from local populations to species and classes. Why is this so? The reason is simply that a peerless adaptedness forged by natural selection in yesterday's and today's environments may turn out to be deficient in future environments. One can hardly imagine a more striking example than the runaway population growth of the human species. It may be the greatest hazard to the adaptedness and even to survival of our species. And yet, the ability of a living population to increase its numbers is, as pointed out above, one of the measures of adaptedness!

Thoday (1953) defined the 'fitness of a unit of evolution' as 'its probability to leave descendants after a given long period of time. Biological progress is increased in such fitness'. He also wrote that 'The fit are those who fit their existing environments and whose descendants will fit future environments.' He is speaking obviously neither of Darwinian fitness nor of adaptedness, but of adaptability. Adaptability is the capacity to become adapted to changes in the environments. The persistence, or durability, of a unit of evolution in time, depends, in the long run, on its adaptability. Some so-called 'living fossils', such as the horseshoe crab, have survived apparently unchanged for long geological epochs, presumably because the ecological niches which they inhabit have also persisted without major changes. But these are exceptions; most organisms evolved to occupy ecological niches different from those of their ancestors. Surely, human environments created by culture and civilization are quite different from those of our pre-human ancestors. When challenged by altered environments, a species with high adaptability evolves genetically (or in the case of man also culturally) to fit into the new environments. A species lacking adaptability becomes extinct or, at best, persists as a relic.

Whence comes adaptability? It might seem that natural selection cannot be its source, because selection is concerned with what is, and not with what will be. This is, however, not the whole story. Every organism has an adaptedness to live in a range of environments. Adaptedness to only a single constant environment would lead quickly to extinction, because environments are not constant. Therefore, every living species is made by natural selection adaptable to the range of the environments which it encounters regularly, or at least at frequent intervals, in its natural habitats. Inhabitants of the temperate zones have to survive winters as well as summers, and of the tropics rainy and dry seasons. A child, a youth, and an adult man live in different social environments and have different roles to play.

Adaptability is obviously of several kinds. There exist physiological and genetic adaptive processes. Physiological homeostasis enables life to go on, and the essential physiological processes to proceed undisturbed, in the face of changes in the environment. Maintenance of constant body temperature and of the pH level of the blood are standard examples. Developmental homeostasis, plasticity, or flexibility are less easily reversible within an individual's lifetime than are physiological homeostatic reactions. The dependence of stature on the nutrition during the growth period is an illustration. The ability to learn, and the consequent flexibility of behaviour, are, however, most important forms of individual adaptability in human societies. Genetic adaptability is due, on a given time level, to the presence of genetic variability and genetic polymorphisms, particularly in populations of sexually reproducing organisms. Diversity of environments can be exploited best by a population that contains diverse genotypes, with optimal adaptedness in different environments. It is less likely that there will appear an all-purpose genotype which will be optimal in all environments.

Physiological and genetic adaptability are certainly not alternative but complementary. Every genotype which occurs regularly in a population must be at least tolerably adapted to some range of environments, but different genotypes may have different ranges of adaptedness. The genetic adaptability in time, i.e. the genetic flexibility when the environment changes in the course of evolution, is probably a function in part of the genetic diversity available on a given time level. Suppose that a population consists of genetically identical individuals, as clones of microorganisms sometimes do. If an environmental change exceeds the tolerance of the genotype, a genetic adaptation can occur only if there appears a mutant with a more favourable tolerance range. A sexual population is likely to contain a variety of genotypes with different optima and tolerances. Some of them may be selected for in new environments. One need not, however, suppose that genotypes adapted to all possible contingencies will always be available ready made. Unless the environmental change is very sudden (in terms of numbers of generations), the selection may gradually compound new genotypes adapted to new environments from genetic building blocks present in the population.

It is sometimes said that a species on a certain time level was 'pre-adapted' to the environment or to the ways of life which it met or chose at a later time level. For example, erect posture developing in our remote ancestors pre-adapted our less remote ancestors to rely for survival on tool-making and tool use; the versatile hands of these less remote ancestors were pre-adapted for driving automobiles and piloting aircrafts today. Now 'pre-adaptation' is not some sort of a prophetic gift of natural selection. Whatever the selection promoted in our ancestors was favoured because it was useful to these ancestors, not because it is even more useful to us today. Erect posture, versatile hands, tool-using and tool-making abilities, evolved not one after the other, but gradually and together. There was a positive feedback between these morphological and psychological traits. The more the life of our ancestors depended on using and making tools, the more natural selection advanced the versatility of the hands which held the tools. Conversely, the more skilful became the hands the better became the tools which they made. In Washburn's (1968) words,

> Human biological abilities are the result of the success of past ways of human life. Through the feedback relation between behaviour and biology, the human gene pool is the result of the behaviours of times past. From the short-term point of view, human biology makes cultures possible. It poses problems and sets limits. But from the long-term evolutionary point of view it was the success of social systems that determined the course of evolution.

A. R. Wallace questioned the power of natural selection to endow 'the savage man' with a brain much superior to that of his ape-like ancestors. What conceivable advantage could men of palaeolithic times derive from mental abilities which in their remote descendants resulted in the inventions of calculus, physics, and metaphysics? The problem would be insoluble if there existed independent genes for each of these abilities. But this is not so – these abilities are outgrowths of the more basic capacities of abstraction, conceptual thinking, and communication by means of symbolic language. These capacities were useful since the beginning of the process of hominization. and their usefulness to all human beings has been increasing ever since. This is not to say that everybody can be an outstanding mathematician or metaphysician. Natural selection has been, and continues to be, a powerful evolutionary agent in the human species precisely because human populations contain ample stores of genetic variants on which the selection works. Though all non-pathological representatives of the human species do possess certain basic abilities, there is still room for individual variation.

## Normalizing natural selection

Teissier (1945) and Schmalhausen (1949) independently pointed out the need to distinguish two kinds of natural selection. One is directional selection (*sélection novatrice*), which changes the gene pool of the population on which it acts, and the other is stabilizing or normalizing selection (*sélection conservatrice*), which keeps the gene pool constant.

In point of fact, normalizing selection is only one of several forms which may act to keep the gene pool of a population constant in composition. Another such form is balancing selection due to heterosis, i.e. to superior fitness of heterozygotes. Others are frequency-dependent selection and selection in populations which inhabit a variety of ecological niches in the same territory.

Mutations occur in man, as they do in all living species. The mutation process is adaptively ambiguous, in the sense that mutations arise regardless of whether they can be useful where and when they appear, or anywhere. In fact, most newly arisen mutants are harmful to their carriers, at least in the environment in which the species usually lives. Uncontrolled accumulation of mutants would therefore lower the adaptedness of the species.

The control is realized through normalizing natural selection. Suppose that the gene $A$ mutates to a recessive state $a$ at a rate $u$ per generation; $u$ is generally small, say of the orders $10^{-5}$ or $10^{-4}$ so that about 1 in 100,000 or one in 10,000 sex cells carry a newly arisen $a$. Suppose further that Darwinian fitness of $AA$ and $Aa$ individuals is 1, and of $aa$ is $1-s$; $s$ is the selection coefficient, which measures the reproductive disadvantage of being $aa$. In every generation some $a$ genes are introduced in the population by mutation, and some are eliminated because of the lower Darwinian fitness of $aa$. An equilibrium state is reached when the numbers of $a$ eliminated by the selection become equal to those arising by mutation. It can be shown that the frequency of a deleterious recessive gene $a$ in a population at equilibrium will be approximately $\sqrt{u/s}$.

Consider the situation when $aa$ individuals are lethal (do not survive) or sterile (do not reproduce). Their Darwinian fitness is 0, and the selection coefficient is 1. If the mutation rate $u$, is 0·0001, the equilibrium frequency of $a$ in the population will be $\sqrt{0·0001} = 0·01$. In other words, about 1 per cent of the sex cells produced in the population will carry $a$, and about 0·01 per cent of the individuals will be homozygous $aa$; some 2 per cent of the individuals will be the heterozygotes $Aa$, healthy carriers of the gene $a$ concealed in heterozygous condition. If a deleterious mutant is dominant to the ancestral condition, the equilibrium frequency of the deleterious mutant gene in the population will be lower, namely $u/s$. For a dominant lethal (such as retinoblastoma without surgical treatment) the value of $s$ is unity. Hence, the number of cases appearing in a population in every generation will be $2u$, twice the mutation rate (because it takes two sex cells to produce an individual, and a dominant mutation in either will cause the disease to appear).

Unfortunately, very few estimates of the Darwinian fitness of human genotypes are available; not unexpectedly, these concern almost entirely hereditary diseases and malformations, because they produce such appreciable changes in fitness that the depression is easily perceptible. The labour and expense needed to obtain fitness estimates for non-pathological traits are so great that few investigators have ventured into this field. The following estimates are taken from the review by Spuhler (1963).

| Trait | Fitness |
|---|---|
| Retinoblastoma (without surgery) | 0 |
| Infantile amaurotic idiocy | 0 |
| Achondroplastic dwarfism | 0·09–0·10 |
| Haemophilia | 0·25–0·33 |
| Dystrophia myotonica | 0·33 |
| Marfan's syndrome | 0·5 |
| Neurofibromatosis: males | 0·41 |
| females | 0·75 |
| Huntington's chorea: males | 0·82 |
| females | 1·25 |

It should be noted that these estimates are valid, even if based on large population samples, only for the places and times where and when they were made. Retinoblastoma is no longer lethal, if the affected eyes are surgically removed or given certain other treatments. The low fitness of achondroplastic dwarfs is in part due to their appearance deviating from the popular canons of sexual attractiveness, making some of them unable to find mates. Since the onset of Huntington's chorea occurs usually in middle age, the Darwinian fitness of the choreics will be higher if marriages and childbearing occur early than if they occur later. Discovery of more effective medical treatments increases the Darwinian fitness of the 'defective' genotypes, and, with unchanged mutation rates, permits these genotypes to accumulate in the populations.

The problem of normalizing selection can also be approached by recording the total incidence of genetically conditioned disability of all kinds found in human populations. The data of Stevenson (1959, 1961) for the populations of Northern Ireland and some districts in England are probably the best available. Between 12 and 15 per cent of pregnancies that continue longer than four weeks end in abortion by the twenty-seventh week, and about 2 per cent end in still births. Miscarriages in earlier stages of pregnancy are not recordable. A considerable, though not exactly known, proportion of the miscarriages and still births are due to genetic defects. About 26·5 per cent of hospital beds are occupied by genetically handicapped persons; 7·9 per cent of consultations with medical specialists and 6·4 per cent of those with general practitioners involve such persons. According to the extensive review by Kennedy (1967) congenital anomalies of various sorts are found the world over in between 1 and 5 per cent of live births, the figures depending largely on the criteria used in the records. During the 1960s a large amount of literature accumulated on the chromosomal aberrations in man. Duplication of chromosome 21 of the normal set gives the so-called Down syndrome, presence of two $X$ chromosomes and a $Y$ chromosome is responsible for the Klinefelter syndrome, and presence of a single $X$ without a $Y$ chromosome for the Turner syndrome. Though they are not lethal, the Darwinian fitness of the afflicted individuals is very low, or zero, because of the sterility. Mental diseases and mental retardations are often genetically conditioned, although the modes of inheritance are mostly unclear or under dispute. Insofar as these infirmities decrease the reproductive success of the individuals affected, they are acted upon by normalizing selection.

## Balancing selection

Some genetic variants cause heterosis, a Darwinian fitness of heterozygotes superior to that of homozygotes. The consequences of this are quite interesting. Suppose that the frequencies of the gene alleles $A_1$ and $A_2$ in a population are $p$ and $q$, and that the fitness of the genotypes is as follows:

| Genotype | $A_1A_1$ | $A_1A_2$ | $A_2A_2$ |
|---|---|---|---|
| Frequency | $p^2$ | $2pq$ | $q^2$ |
| Fitness | $1-s_1$ | $1$ | $1-s_2$ |

It can be shown that heterotic balancing selection will act to retain both alleles in the population, with frequencies at equilibrium $p = s_2/(s_1 + s_2)$ and $q = s_1/(s_1 + s_2)$. This will happen even if one of the alleles causes, when homozygous, a lethal hereditary disease. For example, if $A_1A_1$ dies before reproduction ($s_1 = 1$), and $A_2A_2$ suffers a 10 per cent reduction of fitness ($s_2 = 0·1$), the stable equilibrium frequencies will be $p = 0·09$ and $q = 0·91$; about 0·8 per cent of the fertilizations will have the lethal disease

($p^2 = 0.09^2$). If the heterozygote has only a slight advantage, say, 1 per cent, over both homozygotes ($s_1 = s_2 = 0.01$), the equilibrium will be established at $p = q = 0.5$, and the population will consist of about 50 per cent of heterozygotes and 25 per cent of each of the two homozygotes.

The superior fitness of heterozygotes for a gene which in double dose causes a usually lethal sickle-cell anaemia is most often cited as the example of heterotic balancing selection in man. This should not be taken to mean that this form of selection is rare in human populations, only that selection phenomena in man have been astonishingly little studied. In homozygous condition, *SS*, the gene causes a disease resulting in death, usually at between 3 months and 2 years of age, although a minority survive even to adolescence. The heterozygotes, *Ss*, not only live but are more resistant, relative to the 'normal' homozygotes *ss*, to a form of malaria widespread in the tropics (*falciparum* malaria).

Allison (1964) carried out experiments, infecting volunteer *Ss* and *ss* individuals with this malaria. The results were dramatic; only two of the fifteen *Ss*, and fourteen out of fifteen *ss* individuals contracted the disease. Since the superior Darwinian fitness of *Ss* is a function of the prevalence of *falciparum* malaria in a given territory, the incidence of the gene *S* is highest where this malaria is pandemic (up to 40 per cent of *Ss* persons in the population), while the gene is rare or absent where malaria does not occur. Studies in various parts of Africa indicate that the fitness of *Ss* persons may be as much as 30 per cent higher than that of *ss*; the fitness of *SS* remains close to zero. When a population with a high frequency of *S* moves to a country free of malaria, the superior Darwinian fitness of the heterozygotes disappears, and the incidence of *S* decreases after some generations. This is apparently happening in the United States in populations of African descent. Good, though not quite conclusive, data indicate that several other genetic variants may also be maintained by heterotic balancing selection because of the protection they confer upon their carriers against malaria or other diseases. Here belong the genes for haemoglobins *C* and *E*, the Mediterranean anaemia (thalassaemia), perhaps that for deficiency of the red-cell enzyme, glucose-6-phosphate dehydrogenase, and certain others.

The Darwinian fitness of a given genotype relative to others may evidently change when the environment changes. Reference has been made above to the disappearance of the heterotic advantage in *Ss* heterozygotes in countries free of malaria. The Darwinian fitness may also depend on how frequent are certain genotypes in a given population. A gene may increase the fitness of its carriers when they are a minority, but decrease the fitness when they become a majority. If so, a frequency-dependent balancing selection may work to increase the incidence of a gene when it is rare, but decrease it when it is too common. A balanced equilibrium is reached when the carriers of that gene have neither an advantage nor a disadvantage compared to non-carriers.

No instance of frequency-dependent selection is established in man, although such selection may be quite common. Neel and Schull (1968) give the following imaginative but not implausible example.

A primitive, polygynous society in which each male was highly aggressive might so decimate itself in the struggle for leadership that it was non-viable. At high frequencies of aggressiveness, the non-aggressive male who could keep aloof from the sanguine struggle might stand a better chance of survival and reproduction than the more aggressive, with his chance a function of the amount of aggressiveness in the group. But at low frequencies of the same traits, the aggressive male who assumed leadership (and multiple wives) would be the object of positive selection, and the more passive the group, the greater his reproductive potential. There would thus be an intermediate frequency at which this phenotype (and its genetic basis) would tend to be stabilized.

Diversifying (called disruptive by some authors) selection is related to but not identical with frequency-dependent selection. Any human population and, indeed, any animal or plant population faces not a single environment but a variety of environments. Different variants in the gene pool are likely to possess greatest fitness in these environments. Theoretical and experimental studies of Mather (1955), Thoday (1959), Levins (1968), and others have shown that selection favouring multiple goals may result in the population becoming polymorphic; it will then contain two or several arrays of genotypes having maximum fitness in different sub-environments which the population inhabits. Another possible outcome of selection in diversified environments may be the establishment of 'all-purpose genotypes', which do at least tolerably well under the entire range of the environments.

Diversifying selection may be an important agency in the human species, although only speculative examples of its action can be given. All human societies, and technologically advanced more than primitive ones, have a multiplicity of occupations, tasks, professions, and vocations to be filled if the society is to function properly. How is this need satisfied? It is satisfied in two ways. Any non-pathological human genotype enables its carrier to be trained for the performance of the functions which a given individual elects or is designated to perform. In this sense, *Homo sapiens* has an 'all-purpose genotype' as its species characteristic. It can produce phenotypes suited to many cultural environments in which people may find themselves.

The adaptive flexibility of the human development pattern, particularly as it concerns the processes of learning and acquisition of social competence, is not at all inconsistent with the existence of genetic variability. To put it simply, it is easier to teach some things to some individuals and other things to others. Almost anybody can be trained as a soldier, or an agriculturalist, or a mechanic. But soldiering seems to be more congenial to some, and tilling the soil to others. It is quite likely, though there is no definitive proof of this, that genes contribute towards differentiation of human tastes and preferences.

An even stronger case can be made for genetic conditioning of some special abilities. Outstanding musical, mathematical, artistic, poetic talents, superior prowess in sports or acts of physical endurance, are almost entirely not within the realm of possibilities for most of us. Historical examples are many, in which individuals endowed with such outstanding abilities fared badly at the hands of their more ordinary contemporaries, *les hommes moyens sensuels*. And yet, it is probable that, statistically considered, outstanding achievements in most fields carry now, and have always carried, advantages in survival of the achievers as well as of their families. If this is so, then a variety of genotypes predisposing their carriers towards different occupations and different roles in the society may be favoured by diversifying selection. This selection will tend to establish balanced equilibria; the incidence of the genes, and of the abilities which they are most likely to produce will, in at least a rough way, correspond to what a society needs or admires.

## Directional selection

We have discussed the forms of selection which, given enough time to act in a reasonably unchanging environment, conduce towards genetic steady states. In normalizing selection, the steady state is due to an opposition of a mutation pressure generating variants of low fitness, and a selection removing them from the gene pool. With balancing selection, the steady state is due to an opposition of selection processes acting on different genotypes or in different environments in the same population.

Directional selection is, in a sense, the simplest form of selection, which must have occurred on a grand scale in evolutionary history. A change in the environment may confer upon one gene allele, or a chromosomal variant, a fitness superior to that which it had before the change. Given enough time, the new favoured allele is substituted for, and displaces entirely the previously prevalent allele. The spread of melanic varieties of moths in polluted districts in England (industrial melanism), of insect pests resistant to insecticides, and of variants of rust fungi able to attack the most frequently planted varieties of wheats, are classical examples of directional selection.

Directional selection has certainly operated in the evolution of man and his ancestors. Does it still continue to operate in modern mankind? Geneticists and anthropologists must admit their inability to give even a single well authenticated example of a directional selective process having been observed in operation. The most widely publicized instance in which such selection was supposed to be taking place was the alleged trend towards decreasing intelligence. A higher average fertility of people with lower intelligence, compared to that of more intelligent people (as measured by IQ tests), has been repeatedly found in several studies in different countries. There is also evidence that a considerable fraction of the variance in the IQ scores is genetic. Therefore, one could have reasonably expected to observe a gradual drop in the intelligence averages, at least in the populations for which differential fertility seemingly favouring lower intelligence has been recorded. And yet the famous surveys conducted by the Scottish Council for Research in Education in 1932 and 1947 failed to verify the expectation. There are several possible reasons why the expectation was unfounded, which cannot be discussed adequately in the present article. In brief, the surveys did not include families with no children attending schools; people with severe mental retardations seem to have quite low fertility.

An idea has gained credence, chiefly among social scientists but also among some biologists, that the biological evolution of mankind came to a halt when our species developed culture. Since then, mankind evolves culturally, but is allegedly stable biologically. Confronted with this assertion, a biologist is in an embarrassing position. As stated above, no fully reliable evidence of directional genetic changes has been secured in human populations. Does this prove that man is no longer evolving biologically? It is appropriate to be reminded at this point that Darwin did not claim to have observed natural selection actually taking place. He adduced instead abundant evidence which showed that natural selection must be taking place. As we see this matter at present, there are two necessary and sufficient conditions for natural selection to operate. These are, first, presence of genetic variation of certain morphological, physiological, or psychological traits; and, secondly, this variation affecting the Darwinian fitness of the possessors of these traits.

The environments and the ways of life of people change with extraordinary rapidity, within time intervals of the order of a human lifetime, not to speak of centuries or millennia. True enough, people become adapted to these changes principally by means of cultural rather than genetic transformations. The dichotomy of 'environmental' versus 'genetic' is, however, invalid, and so is the dichotomy of 'cultural' versus 'genetic'. Cultural and genetic changes always were, and continue to be, connected by feedback relationships.

## Genetic death

Consider again the action of normalizing selection which counteracts the accumulation of deleterious mutants in a population. A steady state is achieved when the average number of mutants arising per unit time (e.g. per generation) equals the number eliminated by selection. The elimination has been called by Muller (1950) 'genetic death'. This emotion-laden phrase has taken root in the literature, although a genetic 'death' need not produce a cadaver. For example, a gene for achondroplasia is removed by genetic 'death' when a dwarf carrying this gene fails to find a mate. A sort of genetic 'half death' occurs when a couple of parents has, for genetic reasons, a single child instead of two.

The numbers of genetic deaths that must occur for natural selection to operate were investigated theoretically by Haldane (1937) and by Muller (1950). With normalizing selection at equilibrium, the number of deleterious genes arising by mutation must be equal to the number eliminated. Therefore, the number of genetic deaths that must occur will be equal to the mutation rate (for recessive deleterious mutants), or will be twice the mutation rate (for dominants that are eliminated chiefly in heterozygous carriers). What seems paradoxical at first sight is that the numbers of genetic deaths are independent of the degree of harm a mutant produces, from complete lethals to slight diminutions of fitness. The degree of harm caused by a mutant will, of course, influence its frequency in a population, and also the average number of generations which will elapse between the origin of a mutant and its elimination. The sum total of the genetic deaths will, however, be determined only by mutation frequencies. Even if the mutation rates of genes considered one by one are low (say, of the order of $10^{-5}$), with tens or hundreds of thousands of genes subject to mutation, there will be hecatombs of 'genetic deaths'.

The heterotic balancing selection has to be paid for by still more genetic deaths. If the homozygotes are less fit than the heterozygotes (as in the case of the sickle-cell condition discussed above), there will in every generation take place some genetic deaths among the former. Let the frequencies of the homozygotes $A_1A_1$ ('normal' non-carrier of the sickle-cell gene) and $A_2A_2$ (the anaemic condition) be $p^2$ and $q^2$ respectively, and their selective disadvantages $s_1$ and $s_2$ (see above, p. 51). The heterozygotes, $A_1A_2$, enjoy the highest fitness (in malarial countries). A population consisting entirely of heterozygotes would be the fittest, but such a population produces some homozygotes, $A_1A_1$ and $A_2A_2$, in the progeny. Compared to such an 'ideal' population, a population consisting of the three genotypes at equilibrium will suffer a loss of fitness amounting to $s_1p^2 + s_2q^2$. Suppose that the anaemic homozygote dies before sexual maturity ($s_2 = 1$), and the non-carrier has a 20 per cent disadvantage ($s_1 = 0.2$). The loss of fitness will then be 0.168, almost 17 per cent of 'genetic deaths'!

The problem of genetic loads imposed on a population by genetic polymorphisms maintained by balancing selection has attracted much attention of population geneticists. In recent years, many studies of such polymorphisms have been carried out, especially with the aid of techniques of separation of enzymes and other protein variants by their electrophoretic mobility. These techniques have revealed that a surprisingly high proportion of genes is polymorphic in natural populations. Estimates for human populations are based on the works of Harris (1970) and Lewontin (1967); at least a quarter of the genes examined proved to be polymorphic; an individual is heterozygous for approximately 16 per cent of his genes. Assuming that these estimates are not grossly in error, and assuming that a human sex cell contains some 100,000 genes, an average individual turns out to be heterozygous for some 16,000 genes.

If an appreciable fraction of these polymorphisms is maintained by heterotic balancing selection, the genetic load would seem to be intolerable. Indeed, suppose that the two homozygotes $A_1A_1$ and $A_2A_2$ have fitness only 2 per cent lower than the heterozygote, $A_1A_2$ ($s_1 = s_2 = 0.02$). The loss of fitness caused by this polymorphism will, according to the formula given above, be 1 per cent; the fitness of the population will then be 0.99 of what it would have been if the population consisted solely of heterozygotes. This may not seem very grave; in point of fact, a loss of this magnitude would probably go undetected, given the present state of human population genetics. Suppose, however, that there are 1000 such polymorphisms, and that the losses of fitness which they produce are multiplicative. The fitness of the population will then be $0.99^{1000}$, or about two-thousandths of 1 per cent of that of a population consisting entirely of heterozygotes. If the loss of fitness means increase of genetic deaths, then no human population, and probably no other living population, can sustain such a loss.

A still more serious difficulty was pointed out by Haldane (1957), and following him by Kimura and others (see Crow and Kimura 1970). Adaptive changes by directional selection, i.e. by substitution of gene variants of superior fitness for those of inferior fitness, entails a 'substitutional load' of genetic deaths. This 'load' may, according to the theory, be so heavy that it will slow down very materially the rate of adaptive genetic changes. To understand this paradox, suppose that there arises in some population a novel favourable mutant. Let the mutant be symbolized as a change of a gene $A_1$ to a superior allele $A_2$. The new mutant will be present originally in a single heterozygous individual, $A_1A_2$, in a population in which all other individuals are $A_1A_1$. Since $A_1A_2$ is superior in fitness to $A_1A_1$, the origin of a favourable mutant means that the population suddenly acquires a genetic load of inferior $A_1$ genes. Directional selection will then start working to increase the frequencies of $A_2$ and to decrease those of $A_1$. The process can be viewed as causing many genetic deaths, owing to differential mortality of the carriers of $A_1$.

Evolutionary changes often involve reconstructions of the genetic endowment of a species by substitution of relatively more favourable alleles at many gene loci. A calculus of genetic deaths needed to achieve the substitutions shows that the adaptive improvement can only be achieved very slowly, over many generations, and at the cost of tremendous numbers of genetic deaths. Has 'Haldane's dilemma' driven the theory of evolution by natural selection into a blind alley? Two ways of escape have been suggested.

The more radical one is to suppose that most, or at any rate many, evolutionary changes are 'non-Darwinian' (King and Jukes 1969). It is postulated that many, or most, mutations which occur in the genetic material are adaptively neutral, i.e. do not change the fitness of their carriers either in positive or in negative directions; that the unfixed, polymorphic, genes found in natural populations are mostly neutral variants; and that a majority of the allelic substitutions that occur in evolution are also neutral. Adaptively neutral changes make no genetic loads and cause no genetic deaths. They merely drift in the population gene pool; their frequencies go up or down by chance alone. This drift, or 'random walk', eventually results in loss of some of the variant genes, while others, on the contrary, become more frequent and supplant their competitors. The amplitude of the fluctuations, and the probabilities of loss and fixation, are functions of the population size; they are greater in small than in large populations (Crow and Kimura 1970).

The fundamental assumptions on which the 'non-Darwinian' theory rests are adaptive neutrality of many or most genetic variants, and consequently their irrelevance to adaptive evolution. A critical analysis of these assumptions cannot be given in the present article. It can only be stated that the assumptions lead to construction of mathematical models describing the expected behaviour of genetic variants in populations. Fortunately, the models lead to some predictions that can be tested by observations and experiments on natural and experimental populations. Such tests occupy the attention of many workers at present.

There is another way to escape Haldane's dilemma that must be mentioned here. Sved, Reed, and Bodmer (1967), King (1967) and Wallace (1968, 1970) have considered so-called 'truncation' models of the action of natural selection. To assume, as classical theorists habitually did, that every gene is selected and produces its genetic deaths independently of other genes and of the ecological factors that govern the population size is obviously unrealistic. Of course, a mutant gene which results in a lethal hereditary disease will kill its carrier no matter what other genes the latter may carry. But this 'hard' or 'rigid' selection need not be universal. The selection may also be 'soft' or 'flexible'; it may remove the carriers of some genetic endowments when the population is crowded or exposed to ecological stresses, and let them survive and reproduce when the environment is more permissive. Likewise, the genetic endowments removed by genetic death may be those having combinations of several or many deleterious genes, while the carriers of any one of them without the others may survive. The numbers and kinds of deleterious genes which tip the balance towards survival or towards death are again likely to depend on environmental and ecological situations.

A genetic death will, then, remove from the population not single unfavourable genes but more or less large groups of them. The number of polymorphisms maintained by heterotic balancing selection if that selection acts according to a truncation model is much larger than it can be if each gene is selected independently of the others (see above). Truncation models, like the 'non-Darwinian' ones, lead to some experimentally testable predictions. All that can be said about these predictions here is that investigations aiming to test them are under way.

**Concluding remarks**

Natural selection in mankind is a formidable topic. This review is of necessity brief, incomplete, and perforce superficial. Genetic technicalities have, with few exceptions, been omitted in order to make the review comprehensible to readers who are not at home in population genetics.

The present understanding of how natural selection operates, especially in man, is far from satisfactory. One can infer from circumstantial evidence that various forms of natural selection act on human populations, but only in a few exceptional instances has conclusive direct evidence become available. Normalizing selection is the simplest, and to an evolutionist perhaps the least interesting, form of selection; there is no doubt that it occurs in human populations and yet its quantitative aspects are much less well known than they ought to be. The view that the biological evolution of mankind became arrested when the cultural evolution began is uncritically accepted by many non-biologists. The invalidity of this view is demonstrable on the ground of theoretical inferences but not, it must be admitted, on the basis of concrete observations.

This is a really shocking state of affairs: scientifically and technologically advanced countries have seen fit to expend huge amounts of effort and money to perfect means for self-destruction and to fly to the moon, but not to learn the most basic facts about the state and the possibilities of mankind's own biological endowment. Nevertheless, important advances in our knowledge have been achieved, especially during the last decade or two. They concern mostly matters of conceptualization and theoretical analysis; collection of factual data in human population has progressed on the whole little, because of the expenses involved. We do realize, however, more clearly than in the past that 'natural selection' is a common name for several rather distinct biological processes which play different roles in the evolutionary process. Normalizing selection, different forms of balancing selection, and directional selection may be taking place separately or simultaneously in different populations and at different times. The distinctions between them must always be made in observational and experimental studies.

# 3 Sociology, biology and population control

## A. H. Halsey

(From *Eugenics Review*, vol. 59, No. 3, September 1967.)

There is no more fascinating strand in the intellectual history of the nineteenth and twentieth centuries than that of the relation between the biological and social sciences.

The question now, and the theme of this paper, is whether, and if so on what terms, the intellectual alliance of the mid nineteenth century can be re-established.

We recognize that there is a long road with many pitfalls between the study of geotactic fruit flies and that of the lumpen proletariat (Dobzhansky and Spassky, 1966), or from the phenomenon of population control by grouse on the Scottish moors to an explanation of reduced fertility among some of their predators such as the English upper middle classes in the 1870s (see Banks 1954, and Chapter 5 below by Wynne-Edwards). Nevertheless, our awareness either of these vast distances or of the oversimplifications of earlier attempts to chart them need not prevent us from recognizing that some intellectual pathways already exist, that some scientific linking principles are known and, above all, that appropriate methods of study have been developed which justify the hope of secure, if complicated, interconnection between fields of inquiry which too often exist in isolation.

Professor Bressler (1968) offers us sufficient reasons for the present state of affairs. They are of three kinds: organizational, scientific and ideological. The principal organizational reason is that academic specialization creates vested interests and therefore 'trained incapacity' and intellectual narrowness. The resulting barriers between scholars have been justified and reinforced by the main scientific reason, which is that reduction in general and instinct theory in particular have been shown to be intellectually unsound. And, finally, the organizational and scientific reasons have been powerfully supported by ideological rejection of social Darwinism and politically reactionary doctrines which have been associated with 'the biological approach' to the study of man.

There is, of course, nothing logically necessary about the ideological division – nothing that is inherent in science which compels us to adopt conservative or radical political views. We can say, with Professor Bressler, that science is politically and socially neutral. But ideas and ideology are hard to separate in the real world and we typically have to ask, as we did of Sweden or Spain in the war, 'On which side is it neutral?' Neutrality as such is perhaps not our fundamental problem. Certainly, it would be difficult to fault Professor Bressler's detached and humane demonstration of the absence of *intrinsic* threat to such social values as freedom and equality in biological theories applied to society. The difficulties stem rather from the passions which guide problem selection in the human sciences. There is no escape from the dilemma that passion is at once both a powerful motive towards as well as a potential corrupter of scientific work. The dilemma is necessary. Our only protection is in the preservation of the institutions of free scientific inquiry.

**The autonomy of sociology**
Professor Bressler is eloquent and persuasive in his review of some of the problems on which productive collaboration is possible between sociologists and their colleagues in genetics, biology and ethology. I do not wish to detract from his justifiable enthusiasm which, in any case, I share. Nevertheless we must see to it that our energies are not wasted in swinging a pendulum. Thus, I would stress the validity, especially hard won in sociology, of disciplinary autonomy. There are many far from trivial problems in the social sciences to which the importation of genetic ideas is either unnecessary or even positively misleading. In the case of Ardrey's prediction of the necessary failure of American armed forces in Vietnam by deduction from the territorial imperative, the absurdity is obvious (Ardrey 1966; p. 5) (on the same reasoning the European conquest of the continent of America was impossible and the United States does not exist). Professor Bressler is tempted into the quip that 'insistence in disciplinary integrity and the intense hostility to any form of reductionism may be the academic version of the territorial imperative'. Maybe: analogies can illuminate. But I think we must take more seriously his

reminder of 'Skinner's *caveat* against assuming that similar patterns of social organization (e.g. pecking orders in the barn yard and orders of precedence in medieval Europe) in man and animals arise from similar contingencies'.

More telling still are the examples of sociological explanation of properties of populations which can also perfectly properly be defined genetically. Thus in recent years we have advanced our understanding of the quality of human populations by temporarily jettisoning the genetic concept of a 'pool of ability' and looking at ability as if it were entirely a product of such social forces as economic growth and the expansion of educational provision. This kind of academic autonomy sets limits to the range of soluble problems that can be tackled within the discipline but *within that range* the use of extraneous (in this case genetic) ideas actually impeded understanding. Nor is it certain that disciplinary autonomy can be only a temporary expedient. I would guess that, for example, the theory of conscious, rational and organized behaviour can only be constructed with models of the cost/benefit type used in economics and not from genetic models. In short there will remain, however successfully we collaborate, many human phenomena which we must approach through Professor Bressler's maxim that 'the most parsimonious study of mankind is man'.

These reservations are above all necessary so that we may insist on the advances which are open to us through collaboration across the disciplines. The study of social and genetic determinants of intelligence is an outstanding example of exciting possibilities here, and I shall return briefly to it. I think that there is good ground for hope that, with the extension of work in population genetics and the development of more refined models of social stratification and social mobility by sociologists, we may go a long way towards a viable theory of the sociogenetics of human ability.

## Darwinian population theory and sociological factors

A no less instructive example of all three aspects of the unsatisfactory history of relations between biology and the social sciences – scientific, organizational and ideological – is the study of population size and population control. The general theory of population size, as is well known, was formulated as sociology by Malthus at the turn of the eighteenth and nineteenth centuries (Eversley 1959)[1] and migrated into biology by chance.[2] But the point I want to bring out here is that the migration was imperfect in the sense that a fundamental sociological element of it was left out.

Darwin took over Malthus's notion that potential geometric progression of numbers was a consequence of the fecundity of species. In Chapter 3 of *The Origin of Species,* he wrote, 'In looking at Nature, it is most necessary . . . never to forget that every single organic being may be said to be striving to the utmost to increase in numbers.' He was unsure as to the checks which restrained geometric progression in the real world but in putting forward four main categories of check he included only Malthus's 'positive' or ecological limiting forces and not his 'preventive' or sociological checks. In other words disaster, famine and disease are there as (1) the amount of food, which sets an extreme limit to species increase, (2) predation by other animals, (3) adverse climate, and (4) communicable disease, but not the 'preventive' controls which, despite the moralistic terminology of vice and virtue in which Malthus cast them (and which even today lead us into confusion as to what the Pope really means), are the basis of modern theories of the demography of human populations.

These theories are sociological rather than ecological or biological in emphasis. By sociological I mean to refer to social interactions within a society – that is to systems of mating rather than random mating, systems of communication, convention, prestige, cooperative labour, etc., which are shared by the members of a group. By ecological I mean to refer to external factors in the environment such as climate or the existence of other species which bear upon the form of society. In this sense Malthus's positive checks are ecological and his preventive checks are sociological.

At first glance the student might be tempted to regard Darwin's omission as constructive, if only on the grounds that, because man is uniquely equipped with rationality and forethought, the idea of planned fertility is an irrelevant conception to apply to other species. The premiss of our student would be more or less correct but the inference is seriously misleading. Thus, Professor Wynne-Edwards has pointed out that biologists are now

in the anomalous position of being committed to Darwin's concept that organisms are always striving to increase their numbers and all that follows from this, and at the same time are finding in actual fact that many animals have efficient adaptations for holding their populations down. Some of these populations are limited to a quite low ceiling density, and it can be shown that they are rarely, if ever, exposed to Darwin's checks at all. (Wynne-Edwards 1968, and see below, Chapter 5, p. 89)

All species, including man, are subject to the ultimate check of limited food supply but the proximate or immediate causes of population limitation are normally to be found in sociological rather than biological factors over a very wide range of species.

This is clear enough for human populations. The rate of population growth in industrial countries during the last hundred years falls well within the limits of both fecundity and food supply. Voluntary limitation of fertility is the operative factor. Thus it has been demonstrated by Ansley Coale (1959) that if immortality were added to the birthright of every American citizen, this would have less effect on the long run population size of that country than a mere 20 per cent increase in fertility. Indeed the social and social-psychological factors operate within alarmingly narrow limits in that if married couples in the United States were now to have completed families of 2·5 children on the average then the population would become stable. But if these couples have one more child each and the figure becomes 3·5, then the population would double in forty years. Either of these outcomes is possible. Both patterns exist in identifiable subgroups of the American population (Whelpton *et al.* 1966). Fecundity, and whether it has changed during the period of industrialization, is largely irrelevant. Similarly with food supply. Both total population and GNP *per capita* have risen in the industrial world over the same period: in at least half a dozen countries the rate of increase has been about 20 per cent per decade for two or three generations. Extrapolation on past American experience of fertility and production produces the result that the average American two centuries from now will be thirty-eight times better off (Kuznets 1958).

**Social control of population size in non-human societies**
But what is less widely appreciated is that sociological factors are also usually, as Wynne-Edwards insists, the proximate causes of population control in animals.

It is of the greatest importance in interpreting this insistence on the relevance of a sociological approach to non-human as well as to human societies that we do not fall into the semantic confusion which equates the words social and environmental. We must begin by remembering that every characteristic (apart from the breeding potential of an individual) is *entirely* acquired and that every characteristic is both genetic and environmental in origin. Secondly, we must recognize that a Darwinian evolutionary explanation which invokes genetic factors interacting with environmental forces other than the social system in general or collective planning in particular, may be adequate for its purpose. A good contemporary example is that of the rapid symbiotic evolution of the myxomatosis virus and the rabbit (Burnet 1966). The South American myxo virus was introduced into Northern Victoria in 1950. All but 1 per cent of the rabbits that were bitten by the mosquitoes which carry the virus died within ten days and the immunity of the survivors was not genetically inherited. The survival prospects for rabbits were obviously catastrophic. But this also set no less serious problems for the myxo virus. Once dead the rabbit carcass rots and the virus with it, or else it is eaten by the insusceptible fox or eagle. For the survival of the virus there had to evolve a higher proportion of resistant rabbits so that infection could be securely passed on to other rabbits. Ten years later the adjustment had emerged. Rabbits had become resistant in 20 per cent of cases and the virus was less virulent than its original form. Both organisms had evolved through random errors and accidents or mutations in the DNA that controlled their inheritance, and natural selection had reduced or eliminated genes inimical to survival for the two species in their particular environmental circumstances. Genetic processes are in interaction with environment of this kind and therefore evolution must be presumed to be going on, though usually less dramatically, in all forms of life including man.

But third, which is the present point of emphasis, sociological factors have to be recognized as by no means confined to human populations. Ethological studies of animals since Darwin have demonstrated this repeatedly without their consequence for the structure of Darwinian theory being explicitly challenged. This is the main burden of Wynne-Edwards's thesis. He expresses it in the proposition that 'conventional competition could be the basic cause of social evolution' or alternatively that 'promoting competition under conventional rules for conventional rewards appears to be the central biological function of society' (see below, Chapter 5, and Wynne-Edwards 1962).

Conventional rules governing competition are well known to ethologists: for example, aggressive territorial behaviour, peck orders and the like, establish relations of inclusion and exclusion, dominance and submission without typically involving killing within a species. Wynne-Edwards's example of the Scottish red grouse shows the operative control of a social system which relates numbers to territory and maintains a social hierarchy, and through these controls mortality and fertility.

### The theory of population homeostasis
So far so good, and the utility of a sociology of non-human societies is illustrated. But how is the population level determined over time and, more particularly, how does this happen in the case of man? The theory which Wynne-Edwards accepts and which he extends to human societies, as did Carr-Saunders in an early work (1922), is taken over by analogy from physiology – the principle of population homeostasis. We have already noted, in the case of the myxo virus and Australian rabbits, an illustration of the idea of equilibrium but *not* homeostasis fitted into the context of Darwinian evolutionary theory. We may also note another biological version of homeostasis theory, but one which is untenable. It is familiar to historians of sociology as Spencer's argument against Malthus that among vertebrates 'the degree of fertility varies inversely as the development of the nervous system'. At the highest development of civilization in man the elaboration of mental activity would, according to this argument, produce a natural balance (Eversley 1959). The sociological version of homeostasis theory advanced by Wynne-Edwards and Carr-Saunders is unproven and indeed also untenable in its application to man. Among the criticisms of it recently made by Dr Mary Douglas is the crucially damaging one that it does not work: if it did we would not be worrying about population explosions in Asia and Latin America (see below, Chapter 7).

Douglas also puts forward the criticism of homeostasis theory that it does not seriously consider underpopulation – an omission which enables Wynne-Edwards and Carr-Saunders to treat the actual population as if it were the optimal population. Certainly such an assumption would not fit the circumstances of many human groups, because there are many activities (and not only productive activities) which require a minimum number of participants. She argues that there is much anthropological evidence to suggest that primitive peoples tend towards under-population and that the latent potential for geometric increase in numbers, far from being a threat to the food supply, is in fact not sufficient for such societies to realize the full possibilities of their environment. If, as Douglas points out, this were also true of animal populations it would remove the basis of the problem – i.e. there would be no question of internal social controls if the positive or external checks kept the population level below the point at which it threatened to over-exploit its food resources. Wynne-Edwards does mention such a case – that of fish in the North Sea. But this example of heavy predation is presumably rare.

Among human groups the case is quite different. All known human societies have social controls on mating. Marriage systems are frequently elaborate among primitive peoples.

The function of these social controls is, of course, the same whether the problem of reproductive limitation is set by limits on the food supply or by more complex optima derived from other needs or drives peculiar to the species. In the case of human societies the latter type of factor is more typical and of greater sociological complexity.

Douglas produces four examples to illustrate her thesis that human groups do make attempts to control their populations but that these are more often inspired by concern for scarce social resources (that is, for objects giving status and prestige) than by concern for dwindling basic resources.

## Population quality

Beyond the question of population quantity there lies the question of quality. Where, as among human beings, social control of fertility takes on elaborate forms and directions (through marriage customs, birth control, infanticide, etc.), how do these sociological factors interact with the genetic constitutions of mating popula-tions to produce an evolution in population quality? Here is an intriguing shared problem between genetics and sociology.

There were strong strands in nineteenth century Malthusian thought which deprecated poor laws and social welfare generally as social forces promoting the survival of the unfit and therefore the progressive deterioration of the genetic quality of industrial populations. In the twentieth century we have seen the same kind of reasoning among psychologists. It runs as follows: first, on the assumption that IQ measures genetic potential and, second, given two empirically demonstrated facts – (1) that social class is directly correlated with IQ and (2) that social class is inversely correlated with fertility – it must follow that the intellectual quality of our population is declining. Indeed it must. The results of the Scottish Mental Survey could be explained by a temporary ameliorative set of environmental trends between 1932 and 1947. Meantime, meritocracy only hastens deterioration of the population as a whole.

But the second fact – the inverse relation between status and fertility which has been characteristic of Western populations since the mid nineteenth century – may be changing. Certainly differentials have narrowed. For example, the Growth of American Family Studies show that the relation among Catholics is U-shaped and that there are in general very small differences of expected family size among non-farm wives when they are classified on the basis of husband's income and occupation (Whelpton *et al.* 1966). Indeed, Dr C. O. Carter (1966) has argued that with the extension of family planning over wider sections of the population in this country we may reasonably look forward to the establishment of a positive correlation between intelligence and fertility and therefore to a rising average intellectual quality in the nation. We have here a complicated example of shifting social characteristics in a population which have crucial consequences for its genetic evolution.

Problems like this take us back through Wynne-Edwards and the grouse to the Darwinian evolutionary theory and its assumption that each individual organism constantly strives to increase its progeny in competition with all others. Yet if this were so why do the grouse, for example, not become steadily more aggressive towards rivals within the group by individual selection in favour of aggressive traits and against submissive behaviour? Why do we in fact observe so many instances of sacrifice of individual benefit for group advantage? This is the classic problem of 'genes for altruism' as J. B. S. Haldane phrased it. They do not fit into the pure Darwinian theory. Sir Arthur Keith's group theory of human evolution in a more general form applying to all social animals is now widely accepted. Local mendelian breeding groups and not individuals are the units of evolution. What remains largely unsolved are the mechanisms of inter-group selection.

Genetically the clues lie in the analyses of gene pools, polygenic inheritance, heterosis and bisexuality, which modify the recombination of genes in succeeding generations away from the influence of individual selection. Nevertheless, the net long-run tendency ought to conflict with group survival.

Sociologically, for Wynne-Edwards, the clues lie in social hierarchy: also, I would add, in the further study of mobility between strata and the social patterns of reproduction.

Meanwhile, a challenging puzzle remains.

## Notes

1　I do not want to suggest here that Malthus's ideas were *sui generis*. The attribution of theories to individuals is usually a convenient but distorting device of group memory. David Eversley has shown in his magnificent history of social theories of fertility that Malthus both had significant forerunners and, if carefully read, can be seen to have put forward the major anti-Malthusian as well as the Malthusian theories in his writings as a whole.

2　'In October 1838 – that is, fifteen months after I had begun my systematic inquiry – I happened to read for amusement *Malthus on Population,* and being well prepared to appreciate the struggle for existence which everywhere goes on, from long continued observation of the habits of animals and plants, it at once struck me that under these circumstances favourable variations would tend to be preserved and unfavourable ones to be destroyed. The result would be the formation of new species. Here, then, I had at last got a theory by which to work.' (Francis Darwin, *The Life and Letters of Charles Darwin,* New York, 1887)

# 4  Functional ethology and the human sciences

## N. Tinbergen

(From *Proceedings of the Royal Society of London,* ser. B, vol. 182, 1972, pp. 397–408).

### Adjustability

The extent of individual adjustability is an aspect of the old 'nature–nurture' problem: to what extent is behaviour genetically programmed and to what extent is it further improved by individual modification? I approach the problem from the angle of phenotypic adjustability – its extent and its possible limitations – because this is the practical problem facing Man.

At a time when 'learning' was overstressed, the reaction of ethologists against the *'tabula rasa'* concept of behaviour development was an extremely useful contribution. But the dichotomous classification into innate and learnt behaviour has rather outlived its usefulness for what behaviour students are now, belatedly, doing: analysing the developmental process (Lehrman 1970). They begin to discover (1) by raising animals in different surroundings and (2) by interfering with internal development (done far less often because it is so much more difficult) that, even though the details of the development are extremely complex and insufficiently known, behaviour patterns can be placed, according to their development, on a scale ranging from highly resistant to variations in the environment to highly modifiable. The fact, demonstrated by the ethologists of the 1930s, that many behaviour patterns develop almost perfectly in either grossly deprived situations or even against contrary environmental pressures is now hardly worth stressing (Lorenz 1965, Hinde 1970). What is relevant to my subject is the facts now becoming increasingly clear, that learning is not random, but is often a highly selective type of interaction with the environment.

The work of Thorpe (1961), Marler (1967), Konishi and

Nottebohm (1969) and others on the development of 'song' in some passerine birds can be taken as an example. Song is an elaborate and distinctive motor pattern which functions as a signal in territorial behaviour and mating. When males of the chaffinch are reared without being allowed to hear the song of an adult male, they do not develop the full adult song, but produce a less elaborate 'warble', which has a few but not all the characteristics of the normal song. For the full song to develop young birds must hear, at an early stage, the song of an experienced adult (which, however, they do not reproduce until much later). Such particularly sensitive periods for learning are now known in many instances. In the chaffinch it has further been shown by Thorpe that not all 'teacher' songs are learnt with equal readiness – the birds are biased (i.e. preprogrammed) in favour of learning those songs that show certain characteristics of the natural song of their species.

The oystercatcher provides another clear example of internal control of learning. Young oystercatchers are conditioned to the type of prey that the parents provide, but conditioning to, for example, a mussel shell does not happen in chicks that are merely fed with blobs of mussel flesh in opened shells – it is only those young that receive food, *and* have to chisel it loose from the shell that become conditioned. These are only a few of the many examples accumulating (see, for example, Seligman 1970, Hinde and Hinde 1973) that show that the genetic instructions for the development of behaviour include instructions for phenotypic adaptation – that even learning is not random, but its occurrence, what is learnt and how it is learnt, are prescribed internally within relatively narrow limits, and in addition that these prescriptions are different in different species – each of them is 'programmed for learning' in its own, and adapted way. This expresses itself not only in limitations of what is learnt, but also in a more positive way. A special example of this is exploratory behaviour. This can only be described as behaviour which has the function of creating the opportunities for individual programming. In control of its motivation in its sensitivity for very special external conditions, in its cessation when the environment has become 'familiar', i.e. as soon as salient aspects have been added to the animal's 'knowledge', it is a very beautifully adapted behaviour – adapted to the need to create opportunities for relevant phenotypic adaptation – for maximum success.

An extreme form of exploratory behaviour, known as 'locality study', is shown by some insects. Best known is its application by solitary *Hymenoptera* in 'homing' to the nest site (Tinbergen 1958, 1972). Manning (1956) found that foraging bumblebees learn the position of some individual plants by means of a locality study. But he found, in addition, that such a locality study is made after the discovery of a new *Hypoglossum* plant with a rich nectar supply, but not after the discovery of a new foxglove. When returning to plants of the latter, of which the flowers are visible from a much larger distance, the bumblebee relies on its roaming flight over a large, known area, and on seeing the flower spikes from a distance. Whether or not it makes a locality study therefore depends on surprisingly detailed aspects of the situation.

Resisting temptation to go into more detail, I want to stress the important point that even where adaptedness depends to a large extent on learning, the modifications are themselves internally, ultimately genetically controlled; they are an individual continuation of the process of adaptation, and this supplementary programming varies from species to species; within a species it varies from one developmental stage to the next, from one behaviour system to another, and even from one situation to another. These aspects of the internal control of modifiability are relevant to our own species as well.

### Ancestral man

What lessons can we draw for our own species from these probes into functional ethology? Man is being studied by so many specialized disciplines that it might seem almost impertinent for animal ethologists to put a word in. But the fragmented state of the human sciences justifies and even requires the mobilization of as many relevant disciplines as possible. Ethologists begin to believe that at least some of the methods developed in their science could, if both their power and their limitations are borne in mind, profitably be applied to some important human problems.

At the start I have to emphasize the difficulty to applying to modern man the same functionally oriented, comparative method as we are applying to other animals. This method works well with products of genetic evolution, which still live in the environment to which natural selection has adapted them, and whose behaviour as a consequence is to a large extent constant throughout the species. But because both our behaviour and our environment have changed so much since the cultural evolution began to gather momentum, we are faced with a bewildering variety of anthropogenic modifications – one could say distortions – of environments, and of behaviour systems. Before we could apply the comparative method, we would have to 'peel off' these cultural variations, and reconstruct the behaviour and the environment of our precultural ancestors. Not until we can perform this dual task of reconstruction of environment and behaviour could we sketch a picture of the genetic adaptedness on which our cultural evolution has been superimposed, and which, conversely, has influenced the directions it has taken. To perform this reconstruction, and then to apply the comparative method to man as well, is what the often, but in many respects unjustly, criticized 'naked-apery' is attempting to do. It is a historical exercise, with all its inherent uncertainties.

We do know a little about the environment of early man. Initially derived from a forest-dwelling 'swinger' he has become a bipedal inhabitant of a more open habitat, one richly provided with an under-exploited food supply. Whether or not Hardy's imaginative idea (1960) will receive confirmation that a semi-aquatic phase has also been involved, can be left open, but there seems little doubt that our immediate ancestors have occupied a terrestrial niche – which, as Schaller and Lowther (1969) have so convincingly argued (see also Schaller 1972a, 1972b), early *Homo* could have invaded without much competition.

For the reconstruction of early human behaviour it is only to a very limited extent possible to draw on fossil evidence. It does reveal the early appearance of, for example, bipedal locomotion and a switch from a vegetarian diet to that of a hunter-gatherer, and early cultural developments such as tool-making and the use of fire. But beyond this we must rely on other methods of reconstruction. These are all based on the conviction that cultural evolution has not been random but that it has affected principally those aspects of our ancestral behaviour equipment that were relying most on individual programming, and has changed to a far lesser extent the more internally programmed, more resistant traits, which therefore must be taken to reflect most clearly our ancestral heritage. Evidence of such environment-resistant traits can be expected to come mainly from two sources. Whatever is least variable between cultures, and whatever is least variable within a culture and appears even in spite of environmental pressures in a culturally modified society, is most likely to reveal an ancient, environment-resistant 'deeper structure'.

It has been pointed out by Morris (1967), and in my opinion with justification, that anthropology and ethnology have until recently tended to concentrate more on differences between cultures than, as would be required for our purpose, similarities. As an example of a programme that aims specifically at the study of such similarities I mention recent work of Eibl-Eibesfeldt (1972). It would be far beyond the scope of this paper, and certainly beyond my competence, to try and sketch what intracultural analysis is revealing about the deeper structure of our modern Western behaviour. The sciences most directly confronted with resistant phenomena within our culture, the psychopathological and the educational sciences, are involved in a process of conceptual and semantic fermentation and are in addition split up in innumerable schools. It would be an extremely difficult but also a very important exercise indeed to cut through the barriers that separate such sciences as palaeontology, archaeology, anthropology, and normal and abnormal psychology of both adults and children, and to extract the already available and quite considerable evidence that is relevant to our topic. This would also help to guide future research towards a better understanding of the 'deeper structure', the ancient roots of human behaviour. In this programme, some methods developed in ethology could be of great help, as can be seen from the collection of studies published recently under the editorship of Blurton Jones (1972).

Yet, in spite of the lack of a unified approach, and notwithstanding the fragmented evidence, it is already possible to make a tentative, inspired guess at the ancestral behaviour equipment of our species – a kind of thumbnail sketch. As we go along we can check, as I did with the kittiwake and the oystercatcher, whether such a sketch would make functional sense, as of course it should if we are guessing in the right direction.

Our reconstruction can best start from an assumption that few can doubt – namely, that our bipedal locomotion, the prolonged helplessness of the human infant and its need for extended parental care, as well as our pronounced sexual dimorphism, are old, hardly modified characters. Cross-cultural as well as archaeological-palaeontological evidence suggests that early man lived in relatively small groups – so small that, as in many other primate societies, all individuals must have known each other personally. There is also no doubt that early man has been more of a hunter than his close relatives, certainly more than the surviving apes (Lee and DeVore 1968). The well-established fact that early man was able to kill animals much larger than himself suggests strongly that hunting was, at least on occasion, done in groups. In this respect comparative evidence on other mammalian hunters strengthens our reconstruction: some members of the dog family, the spotted hyaena and among the cats the lion hunt socially, and this allows them to live in part on much larger prey than their solitary relatives (Mech 1966 1970, Schaller 1972a 1972b, Kruuk 1972). In man, the hunting of the larger animals must have been done by the physically stronger males, since the adult females, even though they could carry their infants around, had to be more tied to a secure base (in this respect our dimorphic species differs from, for instance, wolves). In our hunting groups there must have been collaboration, and comparative as well as intracultural evidence suggests that this must have been based on a dominance order – a phenomenon incidentally that entails much more than the word dominance suggests; lower ranking animals do not simply fear their superiors, they also 'respect' them, follow their leadership, and learn from their example (Chance 1967, Kummer 1968). In man, the male is not only a hunter, but also a provider (Washburn and Lancaster 1968) and to a certain extent an educator; this makes it likely that the nuclear family group has at an early stage included the father as well as the mother. Among animals, those species in which both male and female take part in the care of the young have evolved monogamy, and a long-lasting pair-bond with its accessory of falling in love. If, as seems likely, early man was also monogamous (with perhaps incidental bi- or polygamy), the hunting mode of life created a special problem. Hunting large animals (who have large ranges) required long hunting trips and therefore long absences. This would require strong pair-bonding devices. As such, monogamous animal species use various forms of 'extraneous' behaviour systems, such as joint nest building, feeding of the female by the male, mutual preening or grooming, etc. In man, this is where sexual behaviour seems to come in. Coition between partners is recognized as having a strong bond-reinforcing function. In this context the fact that the readiness to mate is far less cyclical in man than in other species should be considered. It is also significant that the use of sexual behaviour or parts of it for other purposes than mere fertilization is widespread among primates (Wickler 1967). In the present primates it is used mainly as a ritualized signal to stabilize the dominance hierarchy: the male mounting act signals superiority, the female posture inferiority, even in non-reproductive encounters among individuals of the same sex. Sex behaviour was therefore so to speak already available for secondary non-reproductive functions. Comparative studies have shown that such dual use of behaviour patterns for both a primary and a secondary function is widespread in the animal kingdom. To the comparative ethologist, the condemnation of 'sex for pleasure' in marital and premarital context alike seems to reveal a lack of biological knowledge; it also ignores the realities of married life even in modern society.

The young of our species had and have a relatively poor non-learnt behaviour repertoire (though undoubtedly richer than has often been assumed (Blurton Jones 1972, McGrew 1972)). The matrix of movements, signals, sensitivity to signals and motivations is, more than in any other species, improved phenotypically. This is ensured not only by social interaction – at first with the mother, then with peers, then with an even wider circle – but also by the young's own, extremely important, exploratory behaviour. As in other primates, this exploratory behaviour blossoms only in the security of maternal supervision, and later in that provided by other friendly individuals. The long development culminates in full incorporation into the adult society, the result not only of relaxation of parent–infant ties but also of active self-assertion by the adolescents – incidentally, the basis for a 'generation gap' which, under the conditions of a vastly accelerated cultural evolution, has now created the need for a *mutual* adjustment between adult and adolescent instead of a mere waiting for the young to conform. Comparisons with group-living primates and wolves, hyaenas and lions, as well as cross- and intra-cultural evidence in man render it further probable that inter-group hostility, particularly between males, and intra-male group friendships are likewise old characteristics. The gist of L. Tiger's interpretations in *Men in groups* (1969) seems to me to be biologically sound. Whether or not this population structure was accompanied by group territories and consequent territorial inter-group hostility can be left open, although in view of what we know of the importance of intricate knowledge of the hunting area, and of the strong tendency of men in many cultures to behave like group-territorial animals, this would seem to be very likely.

It is not necessary here to work out this reconstruction in more detail. As I have said before, it is my belief that it is in outline sound, and also that by a more systematic and more purposeful collaboration of the many relevant sciences, it can be better substantiated, and also elaborated in much more detail.

## Disadaptation and re-adaptation

Already at this stage it is possible to see a little more clearly how drastically and in what respects our new environment differs most from the precultural habitat. Urbanization is perhaps the most striking development. It has carried with it not only crowding, but the formation of very large and in particular anonymous societies, very far removed from the small in-groups of early man. We are also submitted to an enormously increased quantity of input, not only in the form of general, amorphous sensory input in the auditory and visual sphere, but also in the form of quantity of communication through the mass media. The work of countless industrial workers has become extremely monotonous and very far removed from the meaningful and immediately satisfying occupations of the individual craftsman, and certainly from those of the primitive hunter and the primitive agriculturalist. The education of children has changed almost beyond recognition into an extremely demanding training for modern citizenship.

The question that faces the comparative ethologist is: are there signs that this new situation imposes demands on 'human nature' that exceed the limits of its phenotypic adjustability? Are there intolerable pressures, and are there, conversely, gaps, pockets of missing outlets for behaviour patterns that have strong, perhaps compulsive internal determinants? Ethologists believe that there are such signs, and I select three of them for a brief discussion.

Shortly after the Second World War Bowlby (1951 1969) traced back certain disturbances of social behaviour to disruptions of the early phases of affiliation, of bonding between mother and child. Bowlby saw straightforward deprivation of the presence of the mother during longer or shorter periods as the primary cause of a failure in children, first to form personal bonds with the mother, and subsequently of social bonding of any kind. He argued that socialization comes about by a widening of the circle of friends which is only possible if the first personal bond is successfully established. He makes it clear that the young child needs a stable, loving mother or substitute mother, and that modern social conditions often disrupt or even entirely fail to provide for this early phase of socialization. Work on the development of mother–infant relations and of socialization in other mammals, in part inspired by Bowlby's work, gives increasing support to his thesis. Indeed, it becomes very likely that it is not just the presence of a stable mother figure, but an extremely intricate pattern of maternal behaviour that is required. Even very mildly disturbed mothers, such as slightly insecure, or slightly preoccupied working mothers may unintentionally deprive their children. Conversely, over-intrusive, underoccupied mothers may well, through interfering at moments when a child wants to play on its own or with peers, make a child withdraw. There are further signs that, as in other mammals, the affiliation may have to start immediately after birth, and the importance of the ethological studies of mother–infant interactions that are now being made (Harlow and Zimmermann 1959, Spencer-Booth and Hinde 1971) can hardly be overstressed. The paucity of knowledge of these early phases of human life is astonishing, and so, incidentally, is the assertiveness of many theorizers. A number of incidental, at first glance seemingly disconnected observations of family life in man and other mammals, and also in birds, suggest to me that such studies may well reveal even more widespread damage to socialization than Bowlby has pointed out.

As another possible sign of behavioural stress I should like to mention briefly the disorder, or perhaps group of disorders, now generally called early childhood autism, or Kanner's syndrome (Kanner 1943, Bettelheim 1967, O'Gorman 1970, Tinbergen and Tinbergen 1972). There are indications that the recent widespread interest in this serious aberration is due to a real increase of incidence rather than to belated recognition (while its discoverer Leo Kanner found it difficult, in 1943, to obtain sufficient information, there are now, in Britain alone, some 6000 children officially diagnosed as autistic). The syndrome is characterized by a very nearly total lack of socialization, by complete withdrawal from, and even violent rejection of other persons, and by underdevelopment of speech and a number of other skills. A considerable number of autists are damaged for life.

Together with my wife I have compared some aspects of social and socially determined behaviour of normal and autistic children applying methods developed largely in ethology for the analysis of the motivation underlying non-verbal 'expressions of emotions'. We could confirm and elaborate the evidence which had earlier led Hutt and Hutt (1970) to state that many normal children can on occasion show all the components of Kanner's syndrome. By analysing the situations in which this occurs and by studying the forms of therapy which appear to have success, we arrived at the following, tentative conclusions.

1 The distinction between normal and autistic children is far from sharp, and a considerable number of 'normal' children may well be mildly autistic.

2 Motivational analysis indicates that both temporary and permanent autists live in a state of motivational conflict between hyper-anxiety and, as a consequence, frustrated sociality.

3 This conflict can become so severe that the child withdraws from and rejects not only strange persons and environments, but also those that to a normal child become familiar; this can lead to a rejection even of its closest relatives.

4 As a result, socialization is severely hampered, in fact resisted by these children, and as a consequence learning processes that normally form part of socialization, such as the acquisition of overt speech, exploratory behaviour, and certainly learning by social instruction, are likewise impaired.

5 Whereas naturally (i.e. either genetically or 'organically' damaged) timid children are most likely to develop the syndrome, it may well be caused to a much larger extent than is generally recognized by shortcomings in the social environment, to be found particularly in family life, which in certain strata of urbanized society shows certain disruptions.

In other words, we believe that at least some forms of autism are the consequence, and indications of certain forms of increased social stress, and that autism may well be an 'early warning' of harmful effects of the cultural evolution. The phenomenon is certainly in great need of further study, and ethological methods may well help remove the uncertainty and disagreements about its causation, and so ultimately assist in reducing its impact.

Next to these two problems concerning child-rearing and family life, the cultural changes that have imperceptibly, but none the less drastically, influenced educational practices are of at least equal importance. Most of us take institutionalized education for granted, and, for preparing children for the highly specialized parts they will have to play in the modern anthropogenic environment, schooling of some sort is of course indispensable – to think of abolishing schools, as some do, seems totally unrealistic. But it must not be forgotten that schools are a relatively recent cultural phenomenon, and that as such they have to grow with the times. To the ethologist it is clear that we will have to do some hard thinking about both the aims and the methods of education lest we increase, in this sphere too, social stress to beyond what is tolerable.

Socially, of course, school is a good thing because it brings peers together. This would seem to be in harmony with the 'deeper structure' of human social life. It is also a healthy antidote to the isolation that the need for small families and, paradoxically, urbanized living under the crowded but socially lonely conditions of high rise flats, are threatening to force upon us.

But when we compare the educational activities of present-day schools, and the progressive extension of compulsory schooling, with the educational system of contemporary 'primitive' societies, and, by inference, of ancestral man, a few questions arise that deserve much more serious consideration than they are being given, for the differences are much more striking than is generally realized.

The rapid growth of technology in the widest sense requires, of course, that we prepare each generation for playing its part in a society of increasing complexity. But this very speed of change also carries with it the need for each new generation to fit into quite a different, and far more complicated society than that of its parents. Each generation has to learn a great deal more than the previous one. Even we ourselves have seen, in our own life span, a vast quantitative and qualitative increase of what has to be learnt. And this intensified and extended programme has to be met by children who live in a climate of already increased stress and overall input. Modern conditions force us to raise a generation that is at the same time more knowledgeable and optimally flexible and adjustable. Do our educational practices meet these new and perhaps contradictory demands? The question has often been asked, and many educationalists have voiced doubts. In British infant schools and primary schools a great and beneficial revolution is in progress, but even more far-reaching changes may well be needed. Among the many educational innovators and would-be innovators who are expressing opinions on this issue, we might do well to pay most attention to those who base their proposals on close observation on how children learn, and how they fail. With Maria Montessori still as one of the great modern pioneers, most of these students of child development have stressed the need for less imbibing of 'knowledge' and more 'self-activity'.

To the comparative ethologist, this seems to make eminent functional sense. In 'primitive' societies, presumably in this respect more similar to ancestral communities, learning depends partly on exploratory 'play', partly on social imitation, and only partly on deliberate instruction by adults. It seems to me that we have disproportionally increased the part played by social instruction, and that in doing so we are likely to hamper, indeed to stunt and distort development in two ways: we are quite possibly *suppressing* exploratory learning, and we are undoubtedly calling up serious *resistances* against social instruction. This merits a little elaboration.

It is one of the valuable characteristics of our species that the tendency to explore, the sense of curiosity, continues for much longer in the life of the individual than even in the highest primates. Not only the specialized tasks of scientists and technologists, but many other activities in modern society require open-mindedness and an imaginative, exploratory attitude.

The conditions under which exploratory learning flourishes are security, a minimum amount of interference by adults (as distinct from guarding), time and opportunity, and an environment which invites exploration.

I cannot resist relating one little incident which, although one would hardly expect this from the literature on child development, is in my experience representative. A 12-month-old boy, guarded by his aunt and his grandmother, was observed crawling about over a sandy slope which was bare but for isolated rosettes of ragwort and occasional thistle plants. After having moved over many ragwort rosettes without showing any reaction to them, he happened to crawl over a thistle, whose prickly leaves slightly scratched his foot. Giving a barely perceptible start, he crawled on at first, but stopped a second or so later, and looked back over his shoulder. Then, moving slightly back, he rubbed his foot once more over the thistle. Next he turned to the plant, looked at it with intense concentration and moved his hand back and forth over it.

This was followed by a perfect control experiment: he looked round, selected a ragwort rosette and touched that in the same way. After this he touched the thistle once more, and only then did he continue his journey. To ethologists this is only one of many examples of true experimentation in a pre-verbal child; of highly sophisticated exploration.

Yet to all observers of children it must be striking how early their exploratory interest wanes when their school training gets under way. Among the many educators who see a casual connection between this waning and being 'drilled', the perceptive American educationalist John Holt (1970 1971) deserves particular attention because he has put his finger on two fundamental, yet not sufficiently acknowledged effects of school teaching, with its corollaries of continuous testing and giving good marks as rewards and low marks as punishments. He points out that a high proportion of children not only fail to respond, but actively defend themselves against this form of social instruction (and so defeat the teacher) by a mixture of bored rejection and deep-rooted apprehension, by fear of being found wanting. I am convinced that these points deserve close attention, because the tendency, already noticeable in primary school education, of teaching as much as possible by arousing and stimulating exploratory interest rather than by regimented instruction, is not only biologically sound but yields promising results. Nor is this surprising to those who have watched the complete absorption, the perseverance and the patience of children who 'work' in this way and the intense satisfaction they derive from it. This 'wind of change' is also touching secondary education, and will invade higher education, if only because students who have acquired a taste of joint and guided exploration rebel against 'being told', and already demand more exploratory forms of learning. It would seem to me that such a return to a biologically more balanced form of education could also lead to the raising of a type of person which our society needs now. For these needs have been rapidly changing. While a few generations ago we required above all competent professionals, the emphasis may in the future have to be rather on adjustability, open-mindedness, ability to judge, ability to plan far ahead and similar qualities, not only in the leaders, but also in those by whose consent the leaders exert their influence.

My colleagues working in one of the many human sciences may well wonder what has given me the temerity to stray so far outside my home range. The obvious answer would seem that the future of our species is too important to be left to any one group of specialists. The human sciences are, as no one will deny, still very far from a unified discipline. In building such a discipline the collaboration of biologists will be necessary. And a functional ethologist, who is continuously faced with the precariousness of survival in animals, is in a position to see at least some aspects of man's unique position that may not strike students of man as such – and he is extremely alarmed by what he sees.

As I have argued, it is the comparison of the adaptedness of animals with that of man which reveals that our conquest of the environment is causing habitat changes at a pace that genetic evolution cannot possibly match – not even if directed and speeded up by the, morally and practically doubtful, genetic engineering that some have proposed. In order to retain adaptedness we shall have to rely on phenotypic adjustment, and one of the most relevant new insights is that the range of individual adjustability is severely limited, and yet that our educational practices may not exploit to the full the developmental potential that we do possess.

Opinions differ about the imminence of harmful stress phenomena. I have argued that the behaviour student considers it very well possible, indeed likely that we are reaching a point where the 'viability gap' – the gap between what our new habitat requires us to do and what we are actually doing – is becoming so wide that our behavioural adjustability is already now being taxed to the limit. The conclusion seems inescapable that we shall soon be faced with a task of 'bio-engineering' for the purpose of restoring our adaptedness, or rather of re-establishing adaptedness at a new level. The three examples I have mentioned (damage through mother deprivation, autism, and effects of lopsided teaching habits) may well indicate that this task is already now becoming urgent.

It will involve at the same time a restoration of a tolerable environment and the development in ourselves of the highest possible level of flexibility. It will be a task that will require all our resourcefulness, for, as I said, we have no precedent to go by. Briefly, it will amount to no less than *phenocopying, in a very short time, and without paying the tax of massive weeding out of comparative failures, something that has so far been produced only by genetic evolution, which had aeons of time, and which did pay the price of gigantic numbers of errors.*

In this task we cannot possibly succeed unless we know *in concreto* the new pressures that we are creating, and how these pressures could be either met or reduced. And while functional ethology helps us in identifying these pressures, it will be the knowledge of behaviour mechanisms, and of mechanisms of behaviour development that will have to form the basis for whatever engineering will have to be undertaken.

The execution of such an engineering task may at the moment seem to belong in science fiction, but I am convinced that sooner or later it will become a political issue. Knowing what we do about political decision-making I believe that it will be useless to call upon people's altruism or use other arguments of a moral nature. Rather, the scientist will have to point out that the prevention of a breakdown and the building of a new society is a matter of enlightened self-interest, of ensuring survival, health and happiness of the children and grandchildren of all of us – of people we know and love.

No one can say how soon science will be called upon for advice, but if and when that time comes, we shall have to be better prepared than we are now. The main purpose of my paper is, therefore, to urge all sciences concerned with the biology of man to work for an integration of their many and diverse approaches, and to step up the pace of the building of a coherent, comprehensive science of man. In this effort towards integration animal ethology cannot stand aside – indeed I for one believe that, provided it will be given the opportunity for further development, it can render invaluable services.

# PART TWO

*Population*

# 5 Population control and social selection in animals

## V. C. Wynne-Edwards

(From D. V. Glass (ed.), *Genetics*. New York, Russell Sage Foundation and Rockefeller University, 1968, pp. 143–63.)

My chief purpose is to draw attention to the part that social competition plays in natural selection. It is now well established that vertebrates and arthropods of many kinds are capable of regulating their own numbers (Wynne-Edwards 1962).

Some of the mechanisms have long been known, although their functions were not at first understood. They frequently result in expelling whatever population surplus may build up in a locality, and this puts the individuals that have been displaced at greater risk and often results in substantial mortality. Social interaction may affect the admission of recruits to the breeding stock, with the result that some young adults are kept from breeding for a season, or possibly more than one. This means they lose a part, and in some cases the whole, of their chance of contributing anything to posterity.

Social pressures do, in fact, bear unequally on individuals and can condemn a proportion of any given generation to die prematurely whithout issue. The whole conception of population homeostasis is still comparatively young, and its repercussions on theoretical biology are far from fully explored. Here I would like to give an idea of the scale on which social selection takes place, and try to throw a little light on its possible effects as a process contributing to evolution.

## The biological functions of social adaptations
For a century students of evolution have grown up accepting Darwin's ideas on population ecology, as expressed in Chapter 3 of *The Origin of Species*. 'In looking at Nature,' he says there, 'it is most necessary . . . never to forget that every single organic being may be said to be striving to the utmost to increase in numbers.'

He had already remarked that all living species must necessarily have the power to multiply in a geometrical ratio, although most of the time they are prevented from doing so by a variety of checks. He went on to emphasize that 'The causes which check the natural tendency of each species to increase are most obscure'; but everyday experience and common sense suggested to him that they were likely to fall into four main categories. These are the amount of food, which must give the extreme limit to which each species can increase; the serving as prey to other animals; adverse climatic factors; and the effects of communicable disease, especially on crowded animals.

The capacity for geometric increase, curbed by restraining forces, was of course the basis in Darwin's mind for the struggle for existence, and hence for natural selection. Once it had been generally accepted by biologists, there was no further inclination to question or scrutinize its details or to check it against the results of more recent population experiments and field studies. We are now in the anomalous position of being committed to Darwin's concept that organisms are always striving to increase their numbers, and all that follows from this, and at the same time of finding in actual fact that many animals have efficient adaptations for holding their populations down. Some of these populations are limited to a low ceiling density, and it can be shown that they are rarely, if ever, exposed to Darwin's checks.

Modern studies *have* confirmed the expectation that the factor which ultimately limits the density of animal populations is in most cases the food supply. But it is not usually the executive or proximate factor. The immediate cause of population limitation in any given habitat is generally something completely different, such as an individual space requirement. Nevertheless, it does duty in a rough-and-ready way as a mechanism for limiting the total demand for food.

The necessity for making such a substitution, and for having what amounts to an artificial limiting factor instead of a free-for-all scramble for food, is not difficult to understand. If there were no other checks on population increase except the available stock of food, predatory animals – wolves, for instance – would soon increase to the point at which they would begin to diminish their stocks of prey; the fewer the prey became, the more the survivors would be harried. Research on the red deer (*Cervus elaphus*) inhabiting the Hebridean island of Rhum (where there are no longer any wolves) has shown that, during the time that the herd of about 1500 head have been experimentally managed for venison production, it has been possible to kill one-sixth of the stock each year without causing any cumulative change in the size of the herd. The cull has been distributed proportionally over both sexes and all age groups except the calves of the year. We can assume that under primeval conditions a similar annual off-take of something like one-sixth would have been available to the wolves, as the natural predators, and that they could not with impunity take very much more. A kill of one-sixth per annum would allow them to take in an average month one deer in every seventy, or one in 2000 on an average day. Killing more than that would probably not be difficult if there were many wolves, but it would lead to the depletion of the stock and diminishing yields in future years.

Similar conditions can apply to any predator, including man. For example, we understand clearly nowadays the folly of overfishing, although we may not always be willing to accept the restrictions required to prevent it. The danger is not even peculiar to predators: overgrazing by herbivores, leading to down-graded pastures and lowered fertility, is equally familiar in human experience. Apart from domestic animals, there have been cases in which deer and moose have multiplied in the wild and damaged their food resources. This tends to happen in localities where the wolves have recently been exterminated by man. The final result is that the deer themselves suffer.

The real state of the balance between supply and demand, when one is dealing with living food resources, is usually hidden, and quite different from what one might expect. I can illustrate this by reference to the beaver. In an hour or less an adult beaver can cut down an aspen tree that will provide bark and twigs to feed its family, perhaps for several days. The tree has taken possibly twenty years to attain a productive and economical size for consumption, so that to get the maximum food yield from their habitat, beavers should not cut down more than one stem in 7000 on an average day, if they are to provide for a twenty-year rotation. At this rate, their food must always appear to exist in limitless superabundance, even though they are actually consuming the entire annual increment of the poplar forest and there is nothing to spare.

It seems certain that this common need to hold back on the consumption of food is the reason so many animals have become adapted through natural selection to limit their numbers by self-imposed means. To protect such food resources, a ceiling must be put on population density while there is still an apparent abundance of food available. One standing crop must be eked out until the next is ready to be harvested, and demand kept down to the rate at which the supply can be replenished. Such provident ceilings are established by a variety of methods, all of which basically depend on convention. The simplest kind is the subdivision of the habitat and the food it contains into individually held territories. The conventional feature here is that the owner must claim an area big enough to allow him to take all the food he requires without ever running the risk of exceeding its productive capacity.

Through long periods of evolution, many of the density-limiting conventions have grown a great deal more artificial and less direct than this. Gregarious fish, birds, and mammals may hold traditional communal territories, sometimes maintained for many generations, very much like those of primitive human tribes. Within their territories the inhabitants autonomously limit the size of their own flock. In other cases, the feeding grounds are undivided and are shared between social groups. The conventional fabric then depends, for instance, on a pattern of traditional sleeping places, each the property of a different group, or even of a particular individual.

I must make clear that it is by no means necessary to have any overt territorial organization to control population density. For example, guppies in an aquarium, given adequate food, limit their own numbers in relation to the volume of the tank. The mechanism consists simply of eating all the surplus young produced (Breder and Coates 1932).

Under most kinds of conventional limitation, aggressive behaviour intervenes when the acceptable population ceiling is exceeded, and the surplus individuals are driven out. At these times, which ones are to remain within the establishment and which ones are to become outcasts is a question of personal status in a conventional hierarchy, and this is a still more abstract kind of system in the homoeostatic machine. Most aggressive competition is itself conventionalized. Hurtful weapons like teeth and claws are often displayed in threat but are seldom used in mortal combat; superiority can be symbolized in far less barbarous ways than by showing savage weapons. Most territorial birds proclaim their ownership by singing. They seldom have any difficulty in dominating intruders of their own species as long as they hold the psychological edge of being on their own ground.

In conventional competition it is rights that are ultimately at stake. The actual contest is always about some substituted, token situation, but a successful encounter secures for the winners the right to belong and live in the habitat, the right to use its resources, especially food, and in many situations the right to reproduce. A male bird without a territory cannot nest; very likely it will be inhibited from maturing sexually, and it may be prevented from feeding in the habitat. The same applies, in most types of gregarious vertebrates, to any low-ranking member of a hierarchy. Rights are accorded only to those that succeed in qualifying themselves either by winning the required kind of conventional property or, in appropriate circumstances, by attaining a personal standing sufficiently high in the social group. These two immediate objectives of competition, the one concrete and the other abstract, are often inseparably combined; property possession can be one of the main symbols of status, and status differences can decide the day between individuals competing for property.

The most unexpected result of my study of homeostatic population control has been realizing that conventional competition could be the basic cause of social evolution. Society appears to be the organization that provides the medium in which conventional competition can take place. Putting it another way, promoting competition under conventional rules for conventional rewards appears to be the central biological function of society. I have developed this theme on a number of occasions elsewhere, and need now remind you of only two points. The first is not to confuse sociability simply with gregariousness; solitary animals like cats or foxes can and do possess elaborate social organizations. The second is to note how accurately the cap fits when we apply it to human social behaviour. Wherever we look we see brotherhood and cohesion binding the members of a social group to one another and to their homeland, but they are infiltrated by differences of opinion, the emergence of leaders, the desire for recognition and personal status, and for one's own side to win. Society is inherently competitive, and I believe it can be defined biologically as 'an organization of individuals capable of providing conventional competition between its members' (Wynne-Edwards 1962).

### Population homeostasis in the red grouse

I want to turn now to a practical example of social regulation at work in a natural population, and I have chosen the species on which we have been working longest at Aberdeen, the Scottish red grouse (*Lagopus lagopus scoticus*). This valuable gamebird lives between sea level and 1000 metres on open hills and moors covered with a heathy vegetation. The dominant plant is the heather itself, *Calluna vulgaris*, and this is also the staple food of the vegetarian grouse. Much of the year the leaves, buds, flowers, and seeds of heather provide almost 100 per cent of the birds' diet, and at no time less than half of it.

On the better grouse moors, heather appears to cover the ground mile after mile. Measurements show that it actually constitutes between 40 and 75 per cent of the vegetation on the dozen or more areas we have chosen for intensive study during the last ten years. The grouse themselves are moderately large, noisy birds, easily flushed by men and dogs for counting, and not too difficult to catch and mark with visible coloured tabs. Much of our work has depended on studying the behaviour of known individuals. There is a visible dimorphism between the sexes, and some recognizable variation in individual plumage, especially in the colouring of the underparts.

At first sight, it is difficult to believe that food could be the ultimate limiting factor, where heather grows in such profusion. The birds seldom exceed a density of 250 per $km^2$ (one per acre) even under the best summer conditions, and they are often below 50 per $km^2$ (one to five acres). But it has been clearly shown that population densities are in fact closely correlated with the amount of heather cover, with its nutrient status, and with the type of soil on which it grows. Heather is a low, shrubby evergreen with tiny leaves and a life span of up to thirty-five years. It is far more nutritious when young than when old, and for this reason in the spring grouse moors are customarily burned in small strips and patches, on a rotation of twelve to fifteen years. Some older heather is necessary to provide the birds with cover.

Grouse carefully select the parts of the plant they eat, and biochemical studies have shown that what they pick off has a higher nitrogen and phosphorus content than what they leave behind. Comparing heather of the same age on two adjacent moors, one on a very acid granite soil and the other on more base-rich diorite, it has been found that the nitrogen and phosphorus content of leaves is substantially higher on the richer soil, and the same applies to certain other elements, including cobalt. Although the amount and average age of the heather cover is practically the same on the two moors, over a period of five years the one with the richer mineral status has had, on average, just twice the breeding population of the poorer one. A pilot experiment to change the mineral status of 16 hectares of heather by applying nitro-chalk has doubled the grouse-breeding density (from three pairs to six pairs) compared with the initial population and with the untreated control area of 16 hectares alongside, on which there has been no change in numbers.

Direct competition takes place between male grouse for a place to live on the moor. A system of territories held by individual cocks operates most of the year, but in the late summer, when the old birds are moulting, the pattern of holdings quietly lapses. Then, within a few fine mornings about the end of September, quite suddenly and surprisingly, a completely new pattern becomes crystallized and remains substantially unaltered for the next ten to eleven months.

Toward the end of August the young birds are becoming full grown. Family parties of half-a-dozen birds or so are breaking up, and forward young cocks assume a new aggressiveness of behaviour. The conventional time for territorial competition begins about dawn. At this season old cocks recovering from the moult may resume their usual positions and indulge in aggressive vocal display for a while. The rest of the population, mostly females and young birds, lies low until the ritual is over, but after that they are all allowed the freedom of the moor for the rest of the day, feeding very much where they please.

But when the critical period arrives, quite suddenly the young males enter the ranks in territorial dawn display. The result is in the nature of a 'general post'. Usually within one to three days the great annual contest is virtually decided. All the suitable ground is taken up and, as far as the males are concerned, the population finds itself sharply stratified into an establishment of successful territory owners and a residue with lower social status, which can henceforward remain on the moor only on the sufferance of the establishment.

Each morning thereafter, weather permitting, there is an intense display of mutual aggression among the established males, now holding some 2–5 or more hectares of ground apiece. During this time the exact territorial boundaries are hammered out. It lasts only an hour or two at first, and the noncombatants can still ride it out or keep out of the way until it is over. Freedom to come and go is then restored. Not many weeks later, some of the females become associated with the territory holders and are accepted into the establishment, although pair formation still remains fluid.

The rest of the birds, the subordinates, exhibit social differences among themselves. Some are bolder and less easily scared off, and can be found on the moor at all times. Others resign sooner, and become vagrants, not going far, but spending much of their time away from the heather and only returning for an occasional hour or two to feed. These are eliminated soonest and, in fact, some mortality from expulsion probably begins even before the new territorial pattern has emerged.

All unestablished birds have by now been identified *de facto* as surplus, as far as next spring's breeding stock is concerned. There are likely to be a few casualties in the establishment during the winter months, and these gaps will quickly be filled by the topmost unestablished birds, leaving the existing pattern unchanged. The experiment of shooting established males has been repeated a number of times, and has shown that well into the winter there is still a reserve of potential breeders left to fill their places. However, the mortality among the unestablished birds is extremely heavy, and when the spring finally comes, few or none are left.

In an average year, the August population consists of about 37 per cent old birds (comprising the parental age group, born in earlier years) and 63 per cent young (by this time ten to twelve weeks old). It is roughly two and a half times the breeding population of the previous April. If the cycle of numbers brings the breeding stock back to the same level a year later, 63 per cent of the August population will have been eliminated again by the end of the following March. Shooting as a source of mortality can be left out of account, because in practice shooting pressure is never high enough to complete the enormous reduction demanded by natural homeostasis, and experiment shows that the grouse do, in fact, bring their population density down to the same final level whether any are shot or not.

It turns out that almost the whole of this mortality results from social causes. Although the details vary, most typically the population is reduced in two large steps, with a long static interval in between. The first drop comes with the establishment of the territorial system in October, which at once stratifies the population and creates a vagrant class. Within the next few weeks the vagrants suffer substantial losses. Nevertheless, some of them manage to adapt to the new regime, and mortality almost ceases again for several months during the winter. But in February or March the territory owners enter a different phase of aggressive behaviour. Pair formation is consolidated, and both male and female begin to defend their territory, not merely in the early morning, but all day long. Henceforth trespassers are not tolerated, and within a few weeks almost all the remaining surplus birds have been evicted from the moor.

The mortality peaks are easily detected. Corpses are found, sometimes on the moor and sometimes on marginal ground where there is little or no heather. Postmortems show that some of the dead birds are emaciated, and some of them carry a high burden of parasites, especially the nematode *Trichostrongylus*. Feathers left clinging to the heather show where predators have struck their prey down. Although actual figures are, in the nature of things, difficult to obtain, the available data show that there were proportionally seven times as many predator kills of displaced birds as there were of established birds. The dates when peak numbers of birds of prey were counted on the study areas coincided with the steps in population reduction.

Mortality among the established birds is difficult to detect because of the rapidity with which the gaps are filled. The death rate can be estimated only where large numbers of territory owners have been marked with coloured tabs and can be individually identified. The best figure available for mortality in the establishment between November and the following August is 18 per cent.

In the red grouse there is practically no long-distance movement; 12,000 birds have been banded, and 95 per cent of the recoveries have been within 2 kilometers of the place of marking. Except on a local scale, immigration and emigration can therefore be left out of account.

With these data, it is possible to draw up a tentative balance sheet for a standard year, in which it is assumed that the mean population level neither rises nor falls. In practice, grouse numbers do usually fluctuate irregularly from year to year, but this appears to take place without entailing any fundamental differences in the various components that make up the mortality.

The figures given in Table 5·1 must in any case be regarded as only approximate.

Table 5.1 *Average recruitment and loss in a red grouse population*

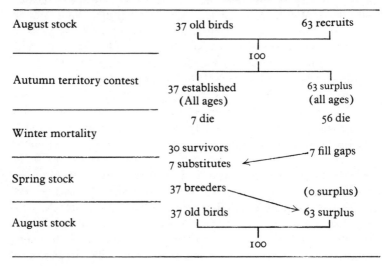

| | | |
|---|---|---|
| August stock | 37 old birds | 63 recruits |
| | 100 | |
| Autumn territory contest | 37 established (All ages) | 63 surplus (all ages) |
| | 7 die | 56 die |
| Winter mortality | 30 survivors | 7 fill gaps |
| | 7 substitutes | |
| Spring stock | 37 breeders | (0 surplus) |
| August stock | 37 old birds | 63 surplus |
| | 100 | |

What the model does, I believe for the first time in a wild population, is to distinguish numerically between the mortality which falls on the socially successful – the territory owners and their mates – and that which falls on the subordinate remainder. The ratio is 7:56, or 18 per cent of the former and 100 per cent of the latter. The two categories owe their distinction ultimately to social competition, so that for grouse beyond the age of 3 months, one can state that social selection accounts for nearly nine-tenths of all the subsequent mortality. Whether young or old, once a bird's social rights have been stripped away it makes little difference from the standpoint of selection what particular agent happens to deliver the *coup de grace*; the bird has already been rejected by the social machine.

To make the recruitment and loss account as complete as possible, I should add that in the embryonic and adolescent phases of life, before the young complete their growth in August, the overall mortality averages close to 80 per cent. Much of this is also likely to be socially induced, but as yet we have not enough information to yield any valid analysis.

## Social fitness

For the red grouse, therefore, success or failure in social competition is undoubtedly a very important component in deciding the 'fitness' of the individual (in the Darwinian sense) because of its influence on survival and reproduction. A bigger surplus is produced in some years than in others, and consequently the pressure of social selection varies, but in our experience of ten seasons on a dozen or more Scottish moors, social competition has emerged as by far the most powerful and consistent selective force. If the environment or food supply suffers a climatic accident, the social organization steps in and influences, at least to some degree, the fate of individuals either as casualties or survivors. If pressure increases from predators or disease, the social status of the individual can influence its relative exposure to risk.

It must be asked whether the red grouse is exceptional in this respect or whether social selection is generally a major phenomenon with far-reaching evolutionary effects.

The beginnings of population homeostasis most probably date to Palaeozoic times. There is an immense span of evolutionary development behind it, and in the most advanced animals it has come to depend on an elaborate complex of adaptations. The evidence indicates that a control system depending on social interaction, in general resembling that of the grouse, is an ancient heritage common to all the classes of vertebrates. The principal features of social organization, including competition for real property and personal status, also can be found in the arthropods. They are less universally obvious there than in the vertebrates, but, particularly in the insects and crustaceans, the signs that indicate social homeostasis are not uncommon. They include territoriality, social breeding and roosting, attachment to traditional localities, and communal epideictic displays. I believe the simplest manifestations of all can be traced to about the evolutionary level of the polychaete worms.

Where social homeostasis occurs, social selection automatically becomes a potential factor in deciding the fitness of the individual organism. But it can take effect only at times when there is reproductive power to spare and a surplus of individuals is produced. At all levels in the animal kingdom some populations are likely to appear in which, for one reason or another, surpluses do not arise. As Darwin pointed out, it may be impossible to establish permanent occupation at the edges of the range of some species because of fluctuations in climate. Frontiers must advance and retreat, and the outermost populations tend to subsist as opportunists and pioneers, at densities that seldom build up to a level that calls for homeostatic controls to be brought into action. In other circumstances, there are species subjected to such heavy and continuous predation that for long periods this is the sole factor determining numbers. In the North Sea, where for a long time the fishing fleets of many countries have exerted a heavy pressure, the stocks of all the main commercial fishes are chronically in this state, and their capacity to reproduce and grow appears to be kept at full stretch. Probably few individual fish in these populations are greatly inhibited, or handicapped in fitness, by social pressures because, with the reduction in population density caused by overfishing, mutual competition must have diminished or even ceased to exist.

A few pioneer experiments reveal the enormous reserve of fecundity possessed by some animals, when subjected to extreme predation stress. A. J. Nicholson (1954) showed that the populations of the Australian sheep blowfly, *Lucilia cuprina*, were not threatened with extinction even though, in one particular experiment, 99 per cent of all the adults were destroyed each day as soon as they emerged from pupation and before they could reproduce. The 1 per cent allowed to live was sufficient to keep a thriving population going and to withstand the 99 per cent predation indefinitely. By the time the experimental population had become completely adjusted, compensatory adaptations had roughly doubled the average lifespan, increased the daily hatch of larvae per female by 50 per cent, and increased the larval survival by a factor of thirty-eight times. What Nicholson calls the effective coefficient of replacement had thus gone up about a hundredfold, and compared with the very much larger control population, they were rearing 5·6 times as many pupae on the same amount of food.

Many studies reveal that the powerful effects exerted by social selection on the red grouse are more or less closely paralleled in other vertebrates. They have been demonstrated in a large number of other birds, in mammals such as primates, rodents, ungulates, and seals, in crocodiles, lizards, and a variety of fish. Although the mortality need not always be equally severe, it seems safe to assume that social selection is a potential force; at least among all the higher animals.

It is important to consider, therefore, the attributes that make for social success or failure in the individual. It can be seen at once that they are not entirely genetic in origin. One of the commonest determinants of status, for example, is the age of the individual. In some species, including many mammals, reptiles, fish, and decapod crustaceans, the males, especially, continue to grow throughout their adult life, so that older adults are automatically stronger than their juniors and enjoy a higher social rank. Alternatively, they may develop progressively finer status symbols in the shape of horns, claws, tusks, manes, crests, and similar adornments. Age differences like these have the function of creating a physical hierarchy in every social group. Often, especially in the male sex, only the senior individuals rank high enough to participate in reproduction. Alternatively, where the qualification for breeding depends on obtaining a property possession, it may be impossible for young contestants to oust the older ones, which have already established themselves in every acceptable site. As I stated earlier, not infrequently this forces young adults to wait, and defers the age at which they first manage to breed, in some cases for several years.

Age effects are tied up with survivorship, and survivorship in turn depends in part on chance, although generally it must also have a partly genetic basis. Individuals that manage to survive long enough to reach the reproductive class will have better adapted genotypes on average than do those that get weeded out prematurely. In organisms that breed for a succession of years, how long an individual stays alive after entering the reproductive class will again, on average, depend in part on its genotype. It appears to be an advantage to the stock to breed as far as possible from such older individuals, which as a class, by surviving longest, have revealed the hereditary qualities that make for adaptability and resistance to injury.

It is clearly desirable that the socially successful individuals should have the qualities that also enable them to survive the real or ultimate agents of natural selection. In an emergency, especially if it entails a shortage of food, the dominant individuals may soon be the only ones left alive, because the rest of the hierarchy has been eliminated by social competition. It seems safe to assume that, however important conventional structures like horns and plumes may be in social competition, they cannot safely be exaggerated to the point at which they impair the chances of individual survival in an emergency. It will be a very important advantage if social success comes easily to the individuals that are best equipped to resist the hostile forces in the environment, both biotic and physical, which from time to time threaten the stock with extinction.

In daily life, an individual's social status is often closely correlated with his physiological condition at the time, but we must use some caution in trying to distinguish between cause and effect. For example, it is not always easy to say whether an individual is dominant because it is well nourished, or well nourished because it is dominant. To give another illustration, subordinate red grouse males are sexually inactive. If in the autumn one of them is implanted experimentally with a pellet of testosterone, he can be stimulated to assert himself and acquire a territory at the expense of established males; consequently he is likely to survive the winter. But once more it is difficult to separate cause and effect, because if an occupied territory is made vacant as a result of the owner being shot in the same season, a similar subordinate male will normally step in, take it over, and become sexually active as a result.

The probability of social success must also depend on whether competition is light or heavy. Where there is a big population surplus, as happens with the lemmings in a peak year, only a minute fraction of dominant individuals will hold their own in the competitive process that commits the vast majority to destruction. In other years there may be no surplus and little or no social competition. Most other kinds of selective forces, including all of those that arise from Darwin's checks, are also variable in their incidence, and there is no reason to think that the varying intensity of social selection nullifies its long-term influence in evolution.

To recapitulate, we have observed that social success can be influenced by the individual's age, by its physiological condition, and by the contemporary intensity of social competition. In human populations another important effect depends on the social status of one's parents, and some trace of a similar prerogative could possibly occur in other species. Assortative mating between individuals of similar appearance has been reported in certain birds, as well as being normal and commonplace in man. But it seems clear that there is generally a major genetic component, in association with secondary and chance effects, that contributes to the social fitness of the individual, and this hands on the results of social selection to succeeding generations.

In our own species, the ability to succeed depends on the sum of the qualities with which an individual is endowed. The qualities themselves can be endlessly varied and combined; some of them are physical, some temperamental or intellectual. There is no one prescription of qualities that can define the successful man. Instead, innumerable different permutations can be rewarded by roughly the same measure of success as judged by the social yardsticks of wealth, influence, and reputation. It is an anomaly that in civilized races people of higher social standing often have smaller families, and in the long run this is disadvantageous; at the barbarian stage of development and among yet more primitive uncivilized races, social distinction and fitness as measured by family size were strongly and positively correlated, as they are in other species of snimals.

I particularly want to draw attention to this non-specific attribute of social selection in man, and to its characteristic property of favouring the individual who can impress his fellows and win their respect as a result of having any one among millions of potentially 'good' combinations of genes.

It appears probable that social selection has a similar basis in the higher animals generally. Like men, rivals of other species confront one another as whole individuals. The better they are matched, the more likely is the conventional issue between them to be decided on points, rather than by a knock out from any one genetic difference. It depends not just on the sharpness of the teeth or the colour of the scales, but on the total effect, which gives the eye its sparkle and spells confidence in action. Genotypes are a mixed bag, and it may not take many shortcomings in one direction to undo the social chances of an individual well endowed in other respects. In the complexity of their social conventions, other animals differ from man only in degree, and probably there are many combinations of genetic ingredients that predispose their owners to social dominance. Thus, social selection appears not to be narrowly channelled and not strongly inclined to restrict genetic variance or to promote the fixation of particular alleles. No doubt feeble combinations of genes tend to be squeezed out in every generation, and those that are good enough to secure for their owners a respectable rating as individuals come through the social mill.

Time does not allow me to develop these ideas much further,

because I have still one other important topic to deal with. It cannot be denied that social selection exists, and that at least in some situations it is a very powerful force. Although social competition relies on a background of real sanctions and the ability to back up postures and threats in actual combat, the artificial symbols of status have become extremely potent weapons. Few hereditary traits appear to be more liable in evolution or more subject to changing fashion than the symbols of personal formidability. In many animals the sexes are dimorphic, and where we have closely related dimorphic species, the most obvious specific differences usually appear in the social insignia of the breeding males. In these cases, population homeostasis and the social organization that promotes it tend to be exclusively male affairs. There is a division of labour, in which the female bears the main burden of reproduction while the male's energies are absorbed in keeping the population density in balance with the available resources; social selection, then, falls more lightly on the female sex. But in the vertebrates it is common to find that, during a long non-breeding period each year, all adults are effectively neuter in sex, their dimorphism is suppressed, and all are competing socially on common terms.

**Inter-group selection**
I want to conclude by touching on one other major aspect of natural selection. This is the effect that arises over long periods of time from the different fortunes of local stocks or populations as entities in their own right: some of them thrive and spread, like human nations in the course of history, while others decline and fall. It is selection acting at the level of the group (Wynne-Edwards 1963).

It arises in part because a social group is much more than merely the sum of its members. Societies have constitutions of their own with regulatory mechanisms, as we have seen, and these elaborate mechanisms can evolve only through a selection that acts directly between homeostatic groups as such. Societies involve a coordination of responses between their members, and their survival is determined by whether they maintain a viable economy. Their efficiency cannot be tested simply by a process of selection acting among their members, any more than the success of a football team can be determined unless the team actually plays against other teams.

As a basis for intergroup selection there must be discrete local populations or groups, each of which is self-perpetuating and capable of maintaining its integrity. No one has done more than Dobzhansky to analyse the kind of spatial and temporal organization that typically exists in animal species. Normally, there is a small flow of genes from one part of the species range to another, usually close by, but it is so slight that the integrity of each local Mendelian group, each with its own gene pool, is not flooded or violated. This characteristic situation is the same as we normally find among underdeveloped human populations.

From the social standpoint, tradition can be important among animals, just as among men, and local stocks often have long-standing customs – for instance, about places for breeding and sleeping, and territorial boundaries. Perhaps nothing shows the importance attached to maintaining discrete local stocks more clearly than the universal acquisition of precise navigational powers by two-way migrants in all the vertebrate classes and even in some insects. These powers allow them to go far away and yet return surely to resume a previously established citizenship.

Evolutionists generally accept the importance of the local population, or deme, as an evolutionary unit (Mayr 1963), and they accept inter-group selection as the process by which can arise, for example, genetic adaptations that affect the group but not the individual. These include the specific number of chromosomes, the frequency of mutations, and the restriction to one sex of crossing-over, as in *Drosophila*. But when it comes to social evolution, doubts have almost always been expressed on the question of selection for self-sacrifice or altruism, which the social code demands. What is best for the social group is often exactly opposite to what is best for the individual. The general conclusion has been, following Haldane (1932), that under natural selection the fitness of the individual is bound to come first: 'genes for altruism', to use Haldane's oversimplified concept, have seemed unlikely to spread.

In effect, evolutionists have been side-stepping this issue, because no alternative explanation has been forthcoming. Societies do exist, and the student of social behaviour is faced every day with real situations in which individual advantage is quite ruthlessly overridden, apparently for the benefit of the group.

The crux of the matter centres on the social hierarchy. The hierarchy is a group adaptation capable of controlling group size and population density; it identifies the surplus individuals and enables them to be got rid of. In vertebrates, as we know, it often tends to allow older individuals to survive, and expends the younger ones. As I have said, older adults may be stronger and better armed, but in many species there is no material difference, and the advantage of age is solely a moral one. Hierarchies often exist even among members of a single age group.

Their essential feature is that some members are always ready to resign themselves to a subordinate rank and be dominated by others, virtually without protest. Social individuals must actually inherit a kind of switch mechanism, which allows them to assume either the dominant or the subordinate role according to circumstances. When animal strangers meet, their switches can be tipped in opposite directions, sometimes in a few moments, just on the basis of general impressions and without any trial of strength. Their mutual rank may be permanently fixed without more ado, and a not uncommon sequel, as we have seen in the red grouse, is for the subordinate to be made an outcast, condemned to an early death.

The pursuit of individual advantage would encourage the loser not to give up so easily, but to return again and again to the attack: it would be better to die in a reckless bid for success by any means, fair or foul, than meekly to resign one's livelihood without even a struggle. Yet we find no evidence of a selection to maximize ferocity toward rivals in the group, or to eliminate submissiveness, any more than there is an inevitable trend toward higher fecundity, or faster growth, or a longer span of life. Each of these characters varies in a given population over a range of values, which must obviously fit the requirements for survival of the group as a whole. The characters are immune to quick selective changes that could give an antisocial hereditary advantage to the individual, because their physiological control is immensely complex and their genetic basis perhaps even more so. Gene recombination in each new generation stabilizes their frequencies near the pre-existing mean. Polygenes and heterosis can negate any immediate hereditary advantage to the progeny of an individual, and if somehow antisocial self-advantage does make a breakthrough and increases the fitness of the individual, it will lead sooner or later to the extinction of the group.

No acceptable alternative inter-group selection has been suggested to explain the evolution of the hierarchic society or of a variety of other more-or-less associated adaptations that characterize the populations of certain species. Most closely related to it is the evolution of castes in the social insects, some of which, being sterile, automatically have no Darwinian fitness at all. Somewhat further removed in function are the sedentary and migratory castes in locusts, aphids and other insects. There is also the evolution and control of polygamy, uneven sex ratios, parthenogenesis, and asexual reproduction. The same selective process has presumably intervened to vary the primary reproductive functions of organisms. Some animals and most of the higher plants are hermaphroditic and every mature individual produces progeny; others are bisexual, with individuals that contribute either sperms or ova but not both, so that the potential fecundity of the population is effectively only half as great.

In evolving all these adaptations, the relative fitness of particular individuals has clearly been more or less irrelevant. It has been subjugated to the requirements for survival of the group as a whole. The facts demonstrate unequivocally that adaptations have arisen, capable of modifying the fitness of the individual in the overriding interests of group survival. To deny this, it seems to me, is to bury one's head in the sand.

# 6 Some economic explanations of fertility

## G. Hawthorn

(Based on a radio talk for the Open University course D101/4, 'Making sense of society', ref. JPM950H212.)

'Economics', an economist once said, 'is all about how people make choices. Sociology is all about why they don't have any choices to make' (Duesenberry 1960, p. 233). He said it in the course of a discussion of a paper on the ways in which couples might choose how many children to have (Becker 1960). That paper proved to be seminal. Since it was published many economists, convinced that family size, at least in industrial societies, is a matter of choice, have speculated about how such choices might be made. Indeed, in so far as there is now anything approaching a developed social theory of human fertility, it is an economic one. It is this theory, or set of theories, that I want to discuss, partly because it is intrinsically interesting, but partly because recent events have begun to cast some doubt upon it.

To begin with, however, it is worth recalling the main steps towards a completed family. This is most easily done by considering them as a series of conditional probabilities (Ryder 1965).[1] First, there is the probability of getting married. Although it is, of course, possible for a woman to have a child without ever marrying at all, most children are still born to married couples and marital fertility therefore accounts for the majority of births. Second, there is the probability of getting married at a particular age. Other things being equal, the earlier a woman marries the longer in the language of demography is she 'exposed to the risk' of conception and the larger therefore is the family that she and her husband are likely to have. Third, there is the probability, once married, of conceiving. Fourth, there is the probability, once having conceived, of the pregnancy coming to term and ending in a live birth. Fifth, there is the probability of the infant, once born, surviving the first year or so, the period in which in all societies death is more likely than at any time before old age. And then, assuming that the marriage continues (and that too, of course, is not certain), there is the probability of conceiving again, and so on round and round the cycle until the woman reaches that point in her late forties at which she is no longer capable of conceiving at all.

The sequence sounds straightforward. As a sequence, it is. What is not, however, is deciding what affects each step in it. If we begin by making a simple distinction between those matters in which a woman (and her husband) can choose, and those in which she cannot, and ignore for a moment the line that Duesenberry drew in this respect between economics and sociology, this becomes clear. How far can one choose whether or not to get married, and when to marry? How far can one choose to conceive, and when to conceive? How far can one choose whether or not to allow a conception to come to term, and if it does, to let the infant survive? And how far can one choose whether or not to interrupt one's marriage? In each case, the answer is: a little, but not entirely. One is no more completely free of physiological capacity, psychological compulsion, economic necessity or social pressure than one is entirely at their mercy. How much choice over what, indeed, is not only a question which it is difficult to answer for any woman, let alone a class of women or a whole society. It is also a difficult question even to ask. There is room for a great deal of disagreement about what may even in principle be meant by and thus count as 'choice'.

I mention this difficulty only to avoid it. The arguments that I shall consider presuppose choice, and I merely wish to make it clear that this is not an easy presupposition to make. Nevertheless, let us assume for the sake of the argument that, once married, couples can choose how many children to have. After all, in healthy and early-marrying populations like this one, they could have eleven or twelve, but most have only two or three.[2]

The argument really began over 2000 years ago (Keyfitz 1972), and the anthropological record suggests that it may have been going on since the Palaeolithic Age (Carr-Saunders 1922, Birdsell 1958, Douglas 1966), but in its modern form it dates from the beginning of classical economics itself at the end of the eighteenth century (Keyfitz 1972, United Nations 1973, Eversley 1959). Since then, it has moved between two assumptions: that children are consumption goods, and that children are factors of domestic production. The first was Malthus's, and until recently it has been the most favoured.

In his famous *First Essay*, Malthus (1970, pp. 70–2) implied that couples would reproduce up to the limits of the available subsistence. Fertility, in modern terms, was a simple positive function of income. By the 1830s, however, he had changed his mind, and was arguing that although this might still be true in principle, in practice, as their incomes increased, couples would prefer and evidently were preferring to consume goods other than children, with the result that the more prosperous might even have a lower fertility (p. 244). In modern terms, again, the implication was that children were an inferior good, like bread. Poorer people wanted more. Richer people wanted less.

The more recent debate about the economics of family formation essentially began with Becker (1960). Becker set himself against the view that children were an inferior good, and argued instead like an early Malthusian that family size was directly related to income. But like Malthus one hundred and thirty or so years before him, Becker had to face the fact that the rich were having fewer children, and not more. He suggested two reasons for this. Richer couples might be preferring to have 'higher quality' children. That is, they might be deciding to spend more on each child than the less rich, and thus, to have fewer children in all. Alternatively, poorer couples might, as the theory predicted, be wanting to have fewer children than the less poor, but their ignorance about contraception or its (for them) relatively high price resulted in them actually having more.

Malthus and Becker both assumed that children were merely consumption goods. But there had long been a suspicion that they also had a productive value. This was reasserted by Leibenstein in 1957, and has since been developed by others, including Becker himself and his colleagues at the University of Chicago. Leibenstein, writing about the process of so-called 'economic development', suggested that in largely agricultural societies, in which most people were living at or near subsistence (perhaps because of being heavily taxed), and in which there was no guaranteed security for old age, children would be valuable both as workers (or, in a market economy, as wage earners) and as a source of support for their parents when they were grown up and the parents too old to look after themselves. In these conditions, he thought, the balance between the costs of children and their value would be far in favour of the second. However, with 'development', with the growth of industry and commerce, with a rise in personal incomes, and with the beginnings of state-financed social security systems, as well as compulsory education, this balance would change. The marginal value of the labour of the $n$th child, as a child and as an adolescent and a young adult, would decline, it would become proportionately less important as a source of financial security in its parents' old age, and it would cost more. Accordingly, it would not be worth having the $n$th child, and it might even be a real disadvantage, materially speaking, to do so. Hence, Leibenstein argued, the transition over time from relatively high fertilities to relatively low ones, and hence too the lower fertility at any one point in time of the rich.[3]

I shall return to the possible contrast in these respects between poor societies and rich ones. But first, I shall explain what has been done for rich societies alone with the assumption that children have a productive value. During the 1960s, Becker and others (Schultz 1974) suggested that households themselves might better be seen not as consumers but as producers. That is, they could be regarded as firms, acting to maximize what in the case of firms are called profits, what in the case of households are most simply seen as household incomes. In the modern labour market, they have argued at Chicago, this means in practice that households will want to realize the largest possible returns on their stock of skills or 'human capital' and their time. The best single measure of human capital is education. In general, more educated people earn more, or can earn more, than less educated people. Thus, if highly educated people, in practice highly educated women, stay at home to look after young children, they are foregoing more potential income, incurring a higher 'opportunity cost', than less well educated women who do the same. Accordingly, and assuming that as members of the family firm all women wish to increase the firm's economic returns, the more highly educated ones will switch their investment away from child-rearing to employment outside the home. And what they have to invest, in addition to their stock of skills, is their time. Hence the emphasis on both education and time. Once again, it is thought, the theory is vindicated by the same basic fact, that the rich, who are also the more highly educated, have fewer children than the rest.[4]

At this point, those not already quite convinced of the validity and the value of such a remorselessly utilitarian model of human behaviour might well object. They might say that people simply do not behave like this, or, at least, that in this part of their lives they do not act rationally in the sense in which a neo-classical micro-economist would interpret that term. Or they might say that even if people are rational in this sense, it does not follow that models of the kind that Malthus, Becker, Leibenstein and others have proposed are valid. Such models suppose a market of explicit trade, but there is no explicit trade in children, at best an implicit or more accurately a quite hypothetical one. Nevertheless, the assumptions are not completely absurd, and since they do imply determinate predictions, these predictions demand to be rationally and empirically assessed. They may even, if only in part, be right.

Much the sharpest and most witty of the briefer assessments has been made by Leibenstein himself (1974).[5] In this paper he could not resist the temptation to suggest a new theory of his own, which he has since developed (1975). But he did have some sharp things to say about the existing ones. Against Becker's earlier view in the 1960 paper that the positive effect of income on fertility might in practice be offset by the rich deciding, as it were, to 'buy' 'higher quality' children, he pointed out that such a view could not account for the facts. The price of higher quality children, more healthy and better educated, was not so high as to offset the propensity to have more. Most industrial societies, except, significantly, the United States, provide health and education at a price far below cost, a price which cannot conceivably make as much difference as Becker suggested to a household's consumption function. And anyway, Leibenstein added, if one actually looks at what parents spend on their children, one finds that richer parents do not spend proportionately that much more. He also pointed out that whatever may have been true about the price of contraception in the 1950s and before, including the price of finding out about it, one could not really maintain that it had not fallen dramatically in the 1960s and 1970s.

Leibenstein then turned to the 'new home economics', as it has been called, to the Chicago theory of the family as firm. In reply to this, he suggested that if more highly educated wives did go out to work and earn the relatively high incomes that the theory presumed, and since they were also married to highly paid husbands, then the family firm could painlessly pay for baby sitting and nursery care and so still have a relatively large, or at least a not unusually small number of children. In short, neither Becker I nor Becker II could satisfactorily account for the inverse relation between income and fertility.

To explain what could, Leibenstein returned to the process of 'development'. In that process, he pointed out, the distance between the highest decile or so of incomes and the lowest decreases. This creates a difficulty for those at the top. They will wish to maintain their privileged social status, and in so far as they can only do so by continuing to consume the sorts of goods that confer status (whatever they may be, but not including children), they will as a result of the compression of all incomes have to spend proportionately more than they had to before, and proportionately more than others, on such goods. In the squeeze, therefore, they switch their spending away from children, and finish with fewer than those below them in the distribution of incomes. This explanation is a more general version of the one that Banks proposed for the fact that in England in the third quarter of the nineteenth century it was the professional and commercial middle classes who were the first to move to smaller families, because they had to find some way of saving in order to continue to be able to pay for wine and servants and all the other goods that set them off, in terms of status, from the aspiring groups below them (Banks 1954). It is both simple and plausible. With it, one now has a third kind of economic theory to explain both falling fertility over time and an inverse relation between income and fertility at one point in time, although a theory which resembles Malthus and Becker I more than it does Becker II in its assumption that children are essentially consumer goods.

Which if any of these theories, however, fits the facts? Consider first the facts of the long-term decline in fertility in the process of 'economic development', say, in Europe since some point in the nineteenth century. Until very recently, there was simply not enough evidence about this decline with which even to begin to test any theory,[6] but this has now changed as the results of a large American research project have begun to appear (Coale 1969 1973). Coale and his colleagues at Princeton decided to collect as much information as they could on the changes in nuptiality (or marriage) and fertility in each county or province or *département* in each European country (including European Russia) from the end of the eighteenth century to the 1930s, and to correlate this information with as much as they could also gather on indicators of social and economic change. They expected to find that fertility at the beginning of the period had generally been at a maximum, and that its decline would everywhere be associated with industrialization, commercialization, a rise in real incomes, the spread of literacy and education, and all the other standard measures of what is sometimes called 'modernization'. The work is not yet finished, but it is already clear for Europe as a whole that fertility was not in general at a maximum at the turn of the eighteenth and nineteenth centuries; and it is clear for Portugal and Germany, the two societies on which work is finished (Livi-Bacci 1971, Knodel 1974), that the decline, dating in the first from the period between 1890 and 1940 and in the second from the unification in 1871, was by no means associated in any simple way with 'modernization'. The best one can say is that the regional differences in fertility before the declines began (there are no data on individuals) do not seem to have been associated with economic differences at all, but with differences of religious tradition and with unidentified local factors of other kinds; and that the start of the declines, and their patterns and paces once they had started, are only very imperfectly associated with economic changes. In 1976, faced with the first results of the Princeton study, it is very difficult to believe in any straightforwardly economic interpretation of the transition to low fertility in Europe.

But what of the present? All the theorists I have mentioned have taken it for granted that what has to be explained is an inverse relation between income and fertility. This relation was indeed characteristic of the period during which fertility was falling (in Britain, say) from the middle of last century to the middle of this. But in Britain, and elsewhere, it has begun to disappear. Already in 1960 in the United States, couples in all income groups were telling interviewers that they wanted the same number of children (Whelpton, Campbell and Patterson 1966, p. 105). And by 1970, again in America, but also in Britain and in other western European countries, rich and poor and middling alike were tending more and more to finish with similar completed family sizes. The very dramatic decline in the yearly birth rate in England and Wales, for instance, since 1965, seems to have been due in good part to the working classes having fewer children and the middle classes having slightly more, the net outcome of which was a decline since there are about twice as many working class couples as there are middle class ones (Office of Population Censuses and Surveys 1974).

Consider what each theorist would now have to argue if he wished to preserve his theory. Becker I would have to say that there was now a coincidental outcome of the rich offsetting their ability to have more children by paying more (although by comparison with their earlier behaviour slightly less) for each one and the poor having to have fewer (but not quite so few) because they were poor. Becker II would have to say that there had been a relative increase in the opportunity cost of the time of less highly educated wives and a relative decline in the shadow price of the time of the more highly educated. And Leibenstein, the Leibenstein of the theory of the consequences of the compression in the income ratio, would have to say that there was now a coincidental outcome of the rich not having to cut back so severely on children for the sake of bigger houses and cars and so forth, for the sake of more status goods, and of the poor having to do so more as their incomes more nearly approached those of the less poor and of the rich themselves. In short, each theorist, to preserve his theory, would have to say that although the kind of choice each couple now makes is the same as it was for the immediately previous generation, the constraints under which each now does so have changed in such a way as to produce a genuine and really rather remarkable coincidence of actual choices and, thus, of outcomes.

We do not have nearly enough information to be able to decide whether or not this is so for any of the three models. But I am sceptical. Relative differences of income remain much as they were two generations ago. It does not seem to be true that more highly educated women are finding it that much more difficult to obtain proportionately well-paid jobs. There is some survey evidence, anyway, to suggest that the working women who want significantly fewer children are not those who work because they want the money but those who work because they want to work, to fulfil themselves outside the home, or just to escape from it (Ryder and Westoff 1971, p. 81). And there is another coincidence, almost certainly a crucially important one, between the recent decline in the birth rate and the convergence of family sizes on the one hand and the spread of reasonably cheap and very effective contraceptives, especially so-called 'oral' contraceptives, on the other.[7] My own view, in fact, is this. Only very recently, in the last decade or two, have societies like Britain entered the final phase of the demographic transition from the pre-industrial period. In terms of Leibenstein's first model, the one that he proposed in 1957, it is only to very recent generations indeed of parents that children really have no value at all as potential sources of income and especially of security, and this, together with the recent and rapid spread of an effective contraceptive (and legitimate abortion), has at last altogether removed children from even the implicit theoretical market. Now, but only now, are parents free to have the number of children that they want for its own sake, no more and no less. As Griliches said (Schultz 1974, p. 547), the Chicago economists had failed to consider family life as an end. Indeed, as far as having children is concerned, it may now almost only be so.

I do not wish, however, to suggest that the economic theory of fertility is finished. Most of the predictions remain still to be properly assessed. But it may well be that the economic determination of fertility is now greater in poor, agricultural societies than in rich, industrial ones. This is somewhat ironic in view of the prevailing prejudice that 'traditional' couples were and are the victims of custom and ignorance whereas 'modern' ones are rational. In poor societies, in general, it is only one's children that stand between oneself and impoverishment in middle life and destitution at the end. The economic theories of fertility may therefore make the most sense where couples do not have any choices to make.[8]

## Notes

1 A fuller account of these events, which does not, however, consider them probabilistically, is Davis and Blake (1956).
2 Estimates of what is usually called 'natural' fertility, the total completed fertility of a woman whose capacity to conceive and to bear children is not deliberately impaired in any way, are notoriously difficult to make. Should one take a hypothetical population, or a real one? If one takes the first, how does one assess the value of the parameters? And if one takes the second, how does one measure the value of these parameters and make sure that no deliberate intervention has occurred? The most comprehensive estimates, in which the assumptions have been made very clear, are in Bourgeois-Pichat (1965). There is a useful summary account in Tabbarah (1971). In practice, many demographers (e.g. Coale 1969) take as their standard the fertility of a population of Hutterite women in the period 1921–30.
3 Leibenstein, of course, neither introduced the notion of a 'demographic transition' nor provides the most complete account of it. For a long summary and two short ones, see respectively United Nations (1973), Coale (1973) and Hawthorn (1970).

4 This argument, with some supposedly corroborating data, was originally proposed in this form by Mincer (1963).

5 There are other useful summaries by Cochrane (1975), Easterlin (1969), Robinson and Horlacher (1971), and Simon (1974).

6 Such evidence as there was, however, was ably summarized and discussed by Wrigley (1969).

7 For the spread of oral contraception, see Ryder and Westoff (1971) on the United States and Cartwright (1976) on England and Wales.

8 This is both paradoxical and contentious. Those wishing to decide for themselves might usefully start with the clear and relatively unprejudiced synopsis in Tabbarah (1971). They could then proceed with Mamdani's (1972) brief criticism of the best existing study of the fertility of a poor population, Wyon and Gordon (1971), consider Mandelbaum (1974) and the more sceptical conclusions arrived at by Macfarlane (1976), by which time they will be well prepared for the exemplary discussion by Cassen (forthcoming). If reasonable, they will conclude that no simple conclusion is in fact possible, although if their conclusions are similar to mine, they will nevertheless remain open to the view that the economic theories do now make more sense in societies like those in rural India.

# 7 Population control in primitive groups

*Mary Douglas*

(From *British Journal of Sociology*, vol. 17, No. 3, September 1966, pp. 263–73.)

This paper is about four human groups which attempt to control fertility. The first are the *Pelly Bay Eskimos* who regularly kill off a proportion of their female babies. The next are the *Rendille*, camel herders in the Kenya highlands. They postpone the age of marriage of women, send numbers of their women to be married to polygamists in the next tribe, kill off boys born on Wednesdays or boys born after the next eldest son is old enough to have been circumcised. The third are the *Tikopia*, inhabitants of a small Pacific island measuring 3 miles across, isolated by 700 miles of sea. They used to use abortion, contraception, infanticide and suicide migration to keep their population down.

These are all groups who by their way of life would be counted as primitive peoples, within the usual range of an anthropologist's interest. I will also mention a fourth group who restrict their numbers by only allowing the eldest son in each family to contract marriage, and correspondingly maintain a large proportion of their female population in barren spinsterhood. These are the *Namburdiri Brahmins* of south India – by no means either poor, or illiterate, or primitive in any sense. I plan to use these examples as a basis for considering Wynne-Edwards's (1962) hypothesis that in primitive human groups social conventions operate homeostatic controls on population.

Wynne-Edwards's thesis is as follows. He asks how a balance is maintained between population density and available resources; what holds back the latent power of increase so that critical resources are not over-exploited? The problem stated thus includes an assumption that the normal distribution of a species is optimum. Wynne-Edwards actually goes so far as to say that normally the habitat provides what he calls 'the best possible living' to species higher up the chain. I quote: 'Where we can still find nature undisturbed by human interference ... there is generally no indication whatever that the habitat is run down or destructively over-taxed. [Each species affords] the best possible living to species higher up the chain that depend on it for food.' His question is about restraint in the midst of plenty. What prevents predators at each point from so multiplying that they over-exploit their own resources?

His answer is inspired by the analogy with homeostasis in physiology. Physiological systems have controls which regulate the internal environment of the body and adapt it. If it can be established that population homeostasis parallels physiological homeostasis, then much behaviour that is apparently functionless can be explained by its contribution to population control. There appear to be density-dependent brakes which impose a ceiling on natural increase. It is important to the argument that the relevant ceiling is *not* imposed by starvation or by predators or natural hazards. Rather it is imposed by otherwise inexplicable aspects of social behaviour. For example:

1 Territorial behaviour limits the number of territories occupied in the food gathering area and deprives redundant males of feeding or breeding facilities.
2 Communal roosting has a function in providing a display of numbers.
3 Hierarchy is a way of cutting off the tail of the population 'at the right level', by excluding certain sections from feeding or breeding.

Finally, the analogy with physiology suggests that the higher species would exhibit more complex adaptations and that population homeostasis would tend to reach the greatest efficiency and perfection in human groups.

Wynne-Edwards extends his argument to human groups by citing enthusiastically a very early work of Professor Carr-Saunders (1922). There is indeed a remarkable close parallel between the approaches of the two authors. In so far as he discusses primitive human populations, Carr-Saunders's argument is as follows.

He starts with the premise that in any human society there will be a theoretical optimum size for the population (that will give the highest return of goods per head). If the density is greater or less than this desirable density, then the average income will be less than it might have been. He goes on to assume that this desirable optimum is actually attained in primitive populations, where it has been observed that the members live in evident enjoyment of satisfactory resources, relatively free from want and disease. To account for this achieved optimum he looks for controls on population. He supposes that starvation is not an acceptable means of limiting numbers, because it makes social conditions unstable, and notes that anyway primitive people are better able to withstand hunger than we are (p. 231). The controls that interest him are imposed from within, social conventions which decrease fertility or increase elimination. Restricted territory, infanticide, head-hunting, such customs are in common use and the degree to which they are practised may be such as to approximate to the optimum number (p. 230). It is only fair to say that Carr-Saunders's book was written a long time ago. None of his other distinguished studies of world population repeats the argument. The anthropological reports which he quotes are out of date and the argument now sounds very naive. I started out with the intention of exposing its fallacies, presented anew by a zoologist. However, if one could adapt the argument to avoid certain inherent difficulties, I would find myself in some measure of agreement with the youthful Carr-Saunders.

The main difficulty with the Wynne-Edwards/Carr-Saunders thesis is that it is so protected from contradictory evidence as to be irrefutable. Wynne-Edwards only expects his thesis to apply where 'nature is kind and reliable'. The negative instances which he cites are said to occur in highly variable environments and so to be compatible with the thesis which is framed for steady environments (p. 470). Carr-Saunders has the corresponding idea that savages are generally found to live in comfort and ease.

These assumptions make it difficult to select relevant data for testing the thesis in its extension to human groups. Are we expected to limit ourselves to savage communities which live in comfort and ease? This could be a very big restriction. What standards of comfort are we to apply, our own or theirs? Peoples whose population is obviously controlled by disease or starvation are to be excluded from the discussion. But there are degrees of starvation. Do we exclude the many peoples who face an annual hungry season between harvests or those who expect a famine every five years, or every ten or twenty years? In short, a principle of selection that conforms to these requirements eludes me. I therefore propose to include any primitive populations for whom good information exists.

The next difficulty with their approach is that underpopulation is not seriously considered. This omission enables them to take the actual given population at any time as the optimum.

If a zoologist tells me that the concept of underpopulation is not relevant to animal groups, I would accept it, but it is highly relevant to human demography because there are many classes of activity which require a minimum number of participants. Much anthropological evidence suggests that primitive populations are prone to *under*population and that the latent power of increase, so far from being a threat to the resources, is not sufficient for the people to realize the full possibilities of their environment. If this were also true of animal populations it would destroy the assumption on which Wynne-Edwards's problem and solution are based. There would be no problem to solve about internal social controls if in fact it could be shown that external controls, in the form of predators and external dangers, kept the populations at each point in the chain down well below the level at which it could threaten to over-exploit its food resources. But this is in fact frequently the case with human groups.

Now for a word about the danger of taking actual human populations at any given time as being at optimum size or density. It is about as defensible as if a town planner were to take the actual size of towns to be the optimum, without analysis or evidence. Thus Carr-Saunders infers from the immensely long prehistoric period through which mankind existed without attaining high densities that some kind of social controls must have operated to produce the optimum size (pp. 239–40). Again, he argues that the existence of restrictive practices such as infanticide implies that the relevant populations are at an optimum size. It is as absurd as for the town planner to infer from the existence of parking meters that the traffic flow is optimal.

The idea of an optimum human population is too complicated to be inferred from such evidence. An optimum density or size can be defined in relation to the demands upon a particular resource. An optimum size in relation to land, for example, would be such that an additional unit of population would not proportionately increase the yield of the land per head, and a subtracted unit of population would more than proportionately increase its yield per head. Such a concept is not always very relevant to actual densities. For example the Ndembu, a Lunda tribe in Zambia, living at a density varying from 3–6 per square mile, grow cassava as their staple crop. Cassava is very easy to grow. It does not require labour-intensive methods. The Ecological Survey of Northern Rhodesia calculated that, cultivating cassava with traditional Ndembu techniques, their tribal area would be capable of supporting a population of up to 18 per square mile. Only near that point of density would the idea of the optimum for cassava cultivation become relevant. At the present density more or less units of human population would not affect the *per capita* yield. In actual fact the Ndembu are not likely to crowd together at the highest densities which their land permits for cassava-growing. Though cassava is their staple, their bread and butter as it were, they are not all that interested in cassava. They are passionately interested in hunting. Game is scarce in their region and the search for it causes them to move their villages when an area is hunted out. It would be nice to think that their actual low density represents the optimum for their hunting economy. But I see no reason for such optimism. They could as likely be *over*populated from the hunting point of view as to have struck a happy equilibrium between their demands on critical resources.

Here is another big difference between human and animal populations when we are thinking of optimum densities. For the animal population it makes sense to make the calculation in terms of critical resources and to recognize that the critical resource is not necessarily food; it may be nesting room or some other necessary amenity. But for human behaviour it can be more relevant to take into account the ceiling imposed by the demand for champagne or private education than the demand for bread and butter.

The shift that has to be made between the zoologist and the sociologist is a shift from the idea of a particular optimum (a size or density related to a particular resource) and a general optimum (a size or density related to the satisfaction of all kinds of demands – including demands for luxuries and leisure).

I give an example of a people who are underpopulated from two angles, economic and social. They are the Western Shoshone, Indians native to eastern California, who used to live by gathering grass seeds and nuts. All the year they wandered from one floristic zone to the next as the seeds and berries ripened, but they wintered near the juniper pine nut crops, wherever they happened to be gathering these when the winter fell. Some of the Shoshone tribes lived at a density of *two to the square mile* in permanent villages. These are the lucky ones. They could sally out for short foraging trips and return to their fixed base. They could get to know each other, hold elections, have winter festivals and organize deer and rabbit drives. Others were much more sparsely scattered at *one person to 2 square miles.* Though they tried to come back to the same place each autumn it was not certain who would be spending the winter together. But at least they could have a festival and could organize a rabbit drive when the others arrived. The least fortunate in the most arid zones were living at a fantastic sparsity of *one person to 30 square miles.* They could never be sure of seeing the same party again from winter to winter, had few festivals and fewer rabbit or deer drives. According to the accepted standards of their own culture they were obviously *under*populated. It is dubious whether these rabbit drives they had to forgo for lack of numbers are to be counted as a critical resource from a strictly economic or physiological point of view. The protein intake of rabbit meat would be very slight and, anyway, their staple was probably not deficient in vegetable protein. The needs which were not met because of low density were social and cultural. But once we admit such resources are relevant to the idea of optimum population we are a long way from both Wynne-Edwards and Carr-Saunders. I am going to argue that it is the demand for oysters and champagne, not for the basic bread and butter, that triggers off social conventions which hold human populations down.

In the absence of any reliable means of calculating a general optimum density I shall take a position close to that implicitly adopted by Carr-Saunders. I shall try to assess what the people living in a particular culture would seem to regard as their optimum size, having regard not only to their demographic policy, but also to the pattern of goals which they appear to set themselves.

Now we come to the final and serious difficulty with the homeostasis theory of human population, which is that it visibly *does not* work. If it did we would not be worrying about a population explosion in India, Mauritius, Egypt, etc.

There are many examples of primitive peoples who hectically

recruit newcomers when their basic resources are visibly running down. The Lele in the Congo were aware of deforestation and erosion, yet each village was more anxious to maintain or increase its *relative* size than to relate size to total resources. Other examples abound of political competition to increase numbers in face of economic pressures to reduce them. What is needed is an account of how population stability is achieved and under what conditions it breaks down. My argument is that human groups do make attempts to control their populations, often successful attempts. But they are more often inspired by concern for scarce social resources, for objects giving status and prestige, than by concern for dwindling basic resources.

Now I am ready to examine the four cases I started with.

First, the Netsilik Eskimos of Pelly Bay – in the 1920s they were an almost isolated group of fifty-four people. Though their area was rich in game their life was one of great hardship, endurance and hard-taxed ingenuity. Rasmussen said that there was scarcely any country on earth so severe and inclement for man. These Eskimo were at a special disadvantage because of their low mobility in winter. They had no driftwood. In the short time between thaw and freeze they travelled by kayak, but in winter their sledges made of old oilskin tents, folded and frozen stiff, or of blocks of ice, were heavy and difficult to move. They kept dogs, not for traction but for locating seals. They went sealing, caribou hunting and fishing in groups or singly, according to the seasonal cycle. Dr Asen Balikci (1966), from whose researches I take this account, considers that in order to survive at all in their environment these people have to show great ingenuity and flexibility. In 1923 Rasmussen noted thirty-eight cases of female infanticide out of ninety-six births for eighteen marriages. Their hunting and fishing economy places great emphasis on the division of labour between the sexes and a man without a woman is at a disadvantage.

Rasmussen was struck by the social difficulties and friction caused by competition for women, often resulting in fighting and killing. He was inclined to argue that the Pelly Bay Eskimo practised female infanticide to a pitch which endangered the survival of the group. However, the more recent anthropologist in the area, Balikci, argues convincingly that the practice is a more flexible and sensitive instrument of demographic policy than was at first supposed. Decisions to kill a new-born infant were taken in the family. If the first-born were female it might possibly be saved, for fear of bringing bad luck on later births, but generally a family was reluctant to take on the charge of rearing a girl, especially if it had a daughter already. A man needed sons to hunt and fish for him when he was past his prime, but rearing a girl would benefit only her future husband. A girl child would not be killed if a future husband would betroth her, or if her grandmother were willing to adopt her as security for old age. So the supply of girls was not simply related to the pressures felt by their own parents. The young men who could not find a wife in their own group had another resource. They could marry girls from another Eskimo group living to the west who did not practise such a high degree of infanticide. Furthermore, although the disparity of the sexes was very marked in infancy, the balance was nearly even for the adult population. The mortality of adult men in hunting incidents, drowning, suicide and fighting, was much greater than for women. Thus, Balikci argues, this group driven to the edge of survival by harsh conditions, in practising female infanticide was contributing to its own survival and making a more or less successful attempt to control the balance of the sexes. Here we have an instance of infanticide genuinely used as an instrument of demographic policy.

According to my general thesis this type of population control in the interests of bare survival is rare. More usually there is prestige rather than subsistence at stake.

The next human group I discuss are the Rendille, a tribe of 6000 camel herders in Kenya (Spencer 1965). The Rendille live on the meat and milk of herds of sheep, goats and camels. They cannot keep cattle because of the aridity and the rough terrain. Camels anyway give two to three times as much milk as a cow, in the wet season, and give adequate supplies in the dry season when a cow gives none. They can survive with water only once every ten days to two weeks, and in the wet season they need no water. They can travel 40 miles a day and so can exploit vegetation in distant areas.

Rendille are aware that their population is limited to the size of herd that can feed it. Each herd requires a minimum number of people to manage it successfully. Smallpox in the 1890s reduced the human population to a too low level of manpower, so they lost stock. The great limitation of camels in these conditions is that the herds increase very slowly. Rendille believe their camels to be a fixed resource. A static stock population cannot support an increasing human population. Rendille are very different from their neighbours, the cattle-owning Samburu, who believe their herds can expand faster than human populations and that a poor man can grow rich in his own lifetime. Rendille have a problem of overpopulation in relation to camels. They deal with it by several measures:

1 By emigration. One third of the Samburu cattle herders are descended from emigrant Rendille, and Rendille still emigrate to this day.
2 By monogamy. A man is not obliged to help his sons to marry a second wife. A herd is not divided and goes only to the eldest son.
3 By late age of marriage for women. A slight excess of women is created by the Rendille custom of monogamy and met by allowing the neighbouring Samburu to marry their female surplus.
4 By killing off boys born on Wednesdays or after the circumcision of the eldest brother, ostensibly to avoid jealousy between brothers.

In this case a shortage of a critical resource, camels, is met with restraints on population. Again, this is a fair case for Wynne-Edwards's general thesis.

If we go on like this, collecting positive instances of successful population control, we finally confront the main question – why do people *not always* practise population control? Why do populations explode? Why do some groups continue to welcome new recruits when crucial resources are visibly running down?

The answer lies in defining more precisely what are the conditions in which a resource is recognized as crucial and limited enough to provoke population policy.

We note that the Rendille camels are in the control of the elders. The whole society is under rigid social constraints, the elders have the whip hand against the juniors, their curse is feared, discipline is tight. In other words, the crucial scarce resource is the basis of all prestige in their society. It happens to be their bread and butter, but at the same time it represents caviar and champagne and all the symbols of status rolled into one.

The next example is the island of Tikopia. In 1929 there were 1300 inhabitants. This group was fully conscious of pressure on resources, as well it might be, 700 miles from the nearest big island and needing to produce all its own food. Strong social disapproval was felt for couples who reared families of more than two, or at most three children. Their population policy was aimed at steady replacement. It was exerted by contraception, abortion and infanticide, and they talked of an ancient custom of pushing out to sea undesirables such as thieves. They lived on fish, root crops (taro and yams) and tree crops (breadfruit and coconuts). Even at the apparently dense population of 1929 they did not seem to feel pressure on land; particularly their rules about lending and borrowing garden land for root crops were very free and easy; they were much more strict about orchard land and particularly coconuts, which produced the cream which made all the other food palatable. Men would fight about orchard land, but not about garden land. In 1952 to 1953 two typhoons in succession produced a famine. Their villages and trees were wrecked and salt spray retarded the growth of their root crops. By this time the population (influenced by missionaries and administration) had relaxed its grip on itself and had increased to 1750. During the famine there were eighty-nine deaths, but only seventeen attributed to starvation. There would have been a higher mortality if relief supplies had not been sent in from the government. The anthropologist, Raymond Firth, who was there in 1929 and also in 1952, gives a fascinating account of the Tikopian reaction to the famines (1939 1957 1959). He considers whether it was famine or fear of famine, which seemed to have occurred every twenty years or so, which actually kept the population down to its size at any given point. But the number of deaths from the 1952 famine is so small, and even those from a 1955 epidemic (only 200) that he did not incline to this Malthusian interpretation. Instead it seems that when they were sedulously restricting their population it was supplies of coconut cream that they had their eye on, not supplies of roots and cereals. Without food of good quality they did not like to hold feasts; without feasts they could not contemplate religious ceremonies; without ceremonies social life came to a standstill. They would exclaim, 'Tikopia does not exist without food ... It is nothing ... There is no life on the island without food.' The anthropologist remarks, 'These expressions alluded not so much to biological survival as to sociological survival'. In making their estimate of how many months it would take to recover from the damage of the first typhoon, when they reckoned that it would be a year at least before the island was on its feet again, there was talk of people putting off to sea in despair. One very old man with experience of previous famines said, 'They say they will die, but they will not die. They will dig for wild yam roots which will not be exhausted and they will go and search for early yams and for wild legume'. Summing up the native attitudes Firth concluded,

> Tikopia did not appear to be concerned with a balance between population and food supply in terms of mere subsistence. They would seem always to have been interested in quality as well as quantity of food, and indeed their estimate of the prosperity of the land is basically affected by this.

My last example illustrates the oysters and champagne factor in population control even more clearly. The Nambudiri Brahmins belong to one of the richest land-owning castes in southern India. They are rich and very exclusive. To maintain their social and economic advantage they avoid dividing their estates, but allow only the eldest son to inherit and to administer it on behalf of his brothers (in the same way as the Rendille camel herders). The other sons are not allowed to marry at all. For each married couple only one son and one girl are likely to be allowed to marry. The other sons console themselves with women of another caste, but the other daughters are kept all their lives in the strictest seclusion (Yalman 1963). Only a very rich community could afford to seclude and condemn to sterility a large proportion of its women, and such a ruthless course must presumably be justified by the value of the prize, in this case maintaining a social and economic hegemony.

To conclude, it seems that population homeostatis does occur in human groups. The kind of relation to resources that is sought is more often a relation to limited social advantages than to resources crucial to survival. In the graded series which I have developed from the hard-pressed Pelly Bay Eskimos to the luxuriously settled Nambudiri Brahmins, the Rendille are important. Their camels are no luxury, but necessary for sheer survival, but I would suggest that the impetus for restrictive policies comes from the great social advantages which accrue to the older men who hold rights in camel herds.

This approach has the possibility of explaining the many cases in which population homeostasis does not appear to work. The argument is that policies of control develop when a smaller family appears to give a relative social advantage. The focus of demographic inquiry should therefore be shifted from subsistence to prestige, and to the relation between the prestige structure and the economic basis of prosperity. A small primitive population which is homogeneously committed to the same pattern of values, and to which the ladders of social status offer a series of worthwhile goals which do not require large families for their attainment, is likely to apply restrictive demographic policies. Such a people would be the ritualistic and feast-loving inhabitants of Tikopia. In a stratified population it is in those sections which are most advantageously placed in relation to power and prestige in which policies of population control are spontaneously applied. Such a people would be the rich and exclusive caste of Nambudiri Brahmins.

When social change occurs so rapidly that the prestige structure is no longer consistent, we should expect population explosions to occur. Or if the whole traditional prestige structure is broken as a result of foreign oppression or economic disaster, again we would expect that the social controls on overpopulation would be relaxed. This happened in Ireland between 1780 and 1840. It is often said that the Irish population made such a remarkable increase in this period because of the adoption of the potato as a cheap form of food. But elsewhere in eighteenth century Europe the potato did not oust other staples (Salaman 1949), and it is unfashionably Malthusian to argue that populations respond directly to increase in the means of subsistence. It is more plausible to adopt my general argument here and to suppose that the ruin of the native Irish society by the penal laws and the ruin of its foreign trade by English tariffs were the cause of the population increase. Similarly, to go further back into English history, the misery caused by the Enclosures and the Poor Laws would have a similar effect and help to produce the manpower for the Industrial Revolution.

It follows that there is a message here for the countries whose prosperity is threatened by uncontrolled population increase. In those countries we see the well educated and well-to-do actively preaching family limitation and setting up birth control clinics as a social service for the teeming poorer classes. They encounter resistance and apathy from the milling poor of the Caribbean, the outcasts of India, the landless labourers of Egypt and Mauritius. Their failure spurs them on to more enthusiastic propaganda. But if they would succeed, let them first look to their prestige structure. What hope of advance does their system of social rewards offer to those to whom they preach? Have the ladders of high prestige enough rungs to reach into the most populous sections of the community? If the prestige structure were adjusted propaganda would be more effective or perhaps not be necessary. For, given the right incentives, some kind of population control would be likely to develop among the poor as it apparently has amongst those who seek to administer the demographic policy.

# 8 Social and economic differentials in fertility

## B. Benjamin

(From J. E. Meade and A. S. Parkes (eds), *Genetic and Environmental Factors in Human Ability*. London, Oliver & Boyd, 1965. pp. 177–84.)

IN the years since the Second World War demography has become a much more developed science than previously. Not only has there been accumulated a wealth of information about the determinants and consequences of population changes and about the character of these changes in populations at different stages of economic development, but demographers have become much better equipped to extract meaning from this information. As a result some early generalizations have been shown to be true not at all times but only at certain times in historical developments, and, in any case, to be oversimplifications. This is especially true of the interrelationship of poverty and large families. While there undoubtedly was a time when large families helped the poor to become poorer and when the poor could command few other interests than procreation, it is no longer so in Britain; nor does it apply to any other society in an advanced state of economic development. It need not necessarily be true of a developing country, since it is at least technically possible with modern contraceptive measures to break the vicious circle of high fertility and a low level of living. Moreover, in speaking of large families the term 'large' has to be related to subjective parental desires rather than to more objective measures such as replacement levels. As will be shown later, we can now generalize more accurately by saying that the lowest social classes tend to have more children than their collectively expressed ideal, while the highest classes tend to come closer to their ideal family size, the ideals differing from class to class. More important still, it will be demonstrated that what happens when fertility is entirely under voluntary control is very different from what happens before this condition is reached. In speaking of fertility differentials in a society, it will therefore be necessary to specify the general extent of voluntary fertility control obtaining in that society.

It will also be necessary to distinguish between economic differentials in fertility at a point of time when 'ideal' family sizes can be regarded as stable for particular social strata, and changes in fertility as associated with economic changes such as trade cycles or sharp rises in unemployment, when 'ideals' may change differently for different social strata.

One important assumption can be made. Since there is no evidence of any real differences in natural fecundity between social groups, it can be assumed that differences in actual fertility must arise from amount of exposure to childbearing, i.e. age at marriage, mortality during the childbearing age, frequency of coitus, and, especially, the degree of practice of contraception.

In Great Britain the first detailed fertility inquiry derived from the 1911 Census (adequate information from vital registration could not be obtained until the passage of the 1938 Population (Statistics) Act). At this census the Registrar General introduced the concept of social classes (eight broad occupational groups). The 1911 Census analysis demonstrated a marked social class gradient in family size (children born). For marriages of completed fertility (wives aged 45 or over at census) the average family size rose from 3·7 in Class 1 (the 'upper and middle class') to 5·0 for Class 3 (skilled workmen) and 5·3 for Class 5 (unskilled workers). A figure of 6·3 was shown for miners. These differences were not greatly affected when account was taken of differences in the age and marriage duration structure of the groups. The census inquiry also showed that the differences between the social classes increased during the latter half of the nineteenth century. It is assumed that this reflected the fact that the knowledge of contraceptive methods became accessible to the better educated and better circumstanced section of society first. It was suspected, and later became clear, that a decline in fertility was then occurring in all classes, particularly in the high fertility classes. We now know that, as a result of legislation against child labour, the latter were finding large families more of an economic burden and, as a result of declining infant mortality, less necessary for the achievement of a desired number of survivors. The gap between the classes was already beginning to narrow.
[1]

When the Royal Commission on Population was set up in 1944 there had been no intervening fertility census since 1911 and the comprehensive vital statistics, collected under the Population (Statistics) Act, dated only from 1938. The Royal Commission therefore set up an organization to carry out a family census, on a voluntary basis, in a 10 per cent sample of married women. In order to examine social and economic differences, women were divided into nine categories according to husband's occupation. For women first married in 1900–9 the number of children born averaged 2·33 for (the wives of) professional workers, 2·64 for employers, 2·37 for salaried workers, 2·89 for non-manual wage earners, 3·96 for manual wage earners and 4·45 for labourers, against an overall average of 3·53. Thus the largest family size was 26 per cent above and the lowest 34 per cent below the overall mean.

For marriages of 1920–4 (a marriage cohort later by fifteen to twenty years) the overall average family size had fallen to 2·42, the spread now extending from salaried employees at 1·65, through professional workers at 1·75 (no longer the smallest), to 1·97 for non-manual wage earners and 3·5 for labourers. Taken at its face value this is a relatively wider spread, but, if one has regard to the two dominant groups of non-manual and manual wage earners, the range for both marriage cohorts remains the same, i.e. from 20 per cent below the all categories mean for non-manual to 12 per cent above for manual wage earners. Because of the shorter durations of more recent marriages it was not possible for the Family Census Report to carry this analysis further in terms of completed fertility, but the statistics for ten-year and shorter durations of marriage suggested that the fall in fertility between the two World Wars had been slightly more marked for the manual group. The gap had narrowed.

The fertility report of the 1951 Census, unfortunately the most recent available, started from this position, i.e. that the fertility differences between social groups increased during the second half of the nineteenth century, were stabilized for several decades and then narrowed for couples married in the 1930s. The census made it possible to bring into the analysis some new dimensions, notably education (or, more exactly, the age of cessation of full-time education) and urbanization, as well as, as in previous analyses, age at marriage. The last two factors had little to contribute to the persisting pattern of family size increasing from Social Class 1 (professional and managerial) to Social Class 5 (unskilled workers), though again the gradient was less steep (85 to 125 per cent of overall average) for women of census age under 50 than for women of census age 45–49 with completed fertility (75 to 132 per cent of overall average). The gap was still tending to narrow.

For the more recent fertility experience Social Class 1 (professional) appeared relatively more fertile, with current fertility equal to the mean for all groups combined. The Report emphasized the significance of the fact that family size in this class, probably the class which started the historic decline in fertility, had apparently begun to rise in relation to the general average, though the available facts, based on a single year preceding the census, were then inconclusive and needed to be confirmed. The trend has not yet been confirmed because the 1961 Census results are not available at the time of writing (1965).

When one looked at the picture in greater detail it appeared that the groups with the largest mean family size and smallest proportion of infertility were the manual workers, especially the unskilled and semi-skilled, members of the armed forces (other ranks), and farmers and agricultural workers. At the other end of the scale were clerical workers, managerial and professional workers, traders and shop assistants.

An important new fact, though not an unexpected one, brought out by the 1951 Census, was that fertility varied inversely with terminal education age, and this variation was slightly more consistent than variation of fertility with socioeconomic group. When standardized for marriage age and socioeconomic group, the index of family size decreased from 101 per cent of the overall mean for those leaving full-time education at ages under 15 to 89 per cent for those pursuing full-time education to ages 20 and over. There were some exceptions, however. For example, for the higher professional administrative and managerial occupations the mean family size of those whose full-time education lasted until after their twentieth birthday was in fact the greatest.

As regards industrial variation the highest fertility was that of wives whose husbands were in the mining and quarrying group, and was associated with their tendency to marry young. Other high fertility groups were those engaged in building and contracting, metal manufacture, and agriculture and forestry.

Married women who were themselves gainfully occupied had lower fertility than those who were not, but of course it is impossible to say which was the cause and which the effect.

There we must leave Great Britain and, for more recent Western experience, turn to the United States of America, where in the last few years there have been many intensive surveys of the growth of families. The USA has shared the experience of some other Western countries, Britain in particular, of an upward fluctuation in numbers of births. In the USA, as here, this has been mainly the result of a trend towards earlier marriage and narrower birth spacing (family-building in shorter time) and, until the marriage cohorts involved have reached longer durations of marriage, it will not be known whether this represents a real increase in fertility as distinct from an alteration in timing. The general impression is that if there is any increase in fertility it is not a trend to larger families but a decrease in childlessness and an increase in families of two, three, and four children. A. A. Campbell (reporting to the World Population Conference 1965 on 'Recent fertility trends in the USA and Canada') has preferred to use educational attainment as an indication of socioeconomic level, because this seldom changes after childbearing has started. He has found that the upward trend in fertility has been strongest in college graduates as compared with those not proceeding beyond high school and has narrowed the gap that previously existed between the two groups. Comparing marriage cohorts of 1901–5 and those of 1926–30, the ratio of family size for college graduates as compared with those who failed to complete elementary school has fallen from 2·4 to 1·4. There has been a change in attitude. Few couples want a childless marriage and the majority at all educational levels want to have two to four children. The change is especially marked in the better educated. This change too has occurred at a time when surveys have shown that 96 per cent of couples with normal reproductive capacity have used or intend to use some method to limit fertility, i.e. when fertility is generally under voluntary control.

Campbell remarks that this last condition means that fertility should be more sensitive to social and economic influences, and he quotes Dudley Kirk (1960) as showing that more couples marry and more try to have children when economic conditions are good. Marriages and births are often postponed when conditions are less favourable. This mainly affects timing rather than the total number of children.

With regard to the question of voluntary fertility control, Westoff (1966) reports that 'while most pregnancies in the USA are unplanned, most couples seem to have the number of children they want', because the effectiveness with which contraception is practised and the proportion practising contraception both increase dramatically as desired family size is approached and achieved. In the USA all socioeconomic groups (and religious groups) are in favour of some control, but family planning attitudes which were previously differentiated by social factors are now converging. Large differences no longer exist in income or occupational groups as such or between urban and rural groups; the main differentiating factors are now education and religion. The less well educated, however, no longer lack information or interest in contraceptive methods. It is true that problems of excess fertility are still severe at the most deprived social levels, but this is the only exception to the general narrowing of social and economic differentials in fertility. With regard to educational differentials, Westoff states that although women of different educational accomplishments are similar in the number of children they desire, they are quite different in the number of children they have had and in the total number they actually expect.

In Japan, another country of advanced economic development, the same trends are found by Kimura (1966). Rapid economic development there has resulted in a decline in the fertility of groups which have hitherto experienced high fertility. This trend to lower fertility has been accompanied by a narrowing of social differentials; in particular, there has been little decline in the fertility of the professional and managerial group, which was already low. In 1955 the standard deviation of family size according to occupation of father was 0·90 (average family size 3·0); in 1960 this had contracted to 0·16 (average family size 2·3). Opinion surveys on fertility show a continuing decline in the rates of childbearing and a convergence among the various social and economic strata, the dominant 'motives' being a desire for higher educational attainment for children, a desire for a higher level of living, and a desire to protect maternal health, in that order of priority.

Generally, then, it appears that, while social and economic fertility differentials persist in developed economies, their intensity is very much reduced as soon as a state is reached when fertility is entirely under voluntary control and when there is much more correspondence between idealized family size, which maximizes educational and social opportunity for the children, and the actual family size achieved. Transient economic changes themselves affect the timing of marriages and births but apparently do not affect long-term desires or their eventual achievement. We can now speak of fertility much more in terms of people's desires and motives. Whether the educational bias in this motivation means that the social differences in fertility, which are now narrowing, may actually be reversed, we do not yet know.

## Convergence and choice

In the aftermath of the Second World War, the so-called 'population explosion' (not really an explosion but not less serious for that) accompanied the emergence of developing nations. Significant population changes have been expected, especially in the already developed countries which, in a position of fertility control (in the sense that individuals may make active rather than passive decisions to procreate or not), were able to react effectively to world events. An authoritative review of this reaction had been overdue. The first and most detailed such review appeared in 1975 as Part II of the *Economic Survey of Europe in 1974 (Post-war Demographic Trends in Europe and the Outlook until the Year 2000)* and published by the UN Economic Commission for Europe. This was a remarkable document – remarkable for its authority, its breadth of coverage both in terms of the completeness of its inclusion of European countries, even of eastern Europe where population and economic statistics have not always been easy to obtain. But this document was necessarily somewhat heavy and difficult to digest; what was needed was a shorter review, scholarly yet articulate, concentrating on the main demographic element – fertility. This came early in 1976 when David Glass was invited to give a Review Lecture to the Royal Society on the subject of 'Recent and Prospective Trends in Fertility in Developed Countries'. This went beyond the Economic Commission for Europe study in that it covered developed countries worldwide and not only in Europe.

The main theme of the lecture was convergence; the diminishing divergence of family-building behaviour as between the populations of different countries; and as between individual couples within these populations. The main idea that it appears to provoke is the possibility of national control of population growth without interfering with real, as distinct from socially engendered, fertility aspirations.

The Second World War marked the end of a demographic epoch. The marriage pattern of west and central Europe – a high age at first marriage and substantial proportions of men and women (especially women) remaining unmarried – gave way to the more natural pattern of low age at marriage and a high (90 per cent) probability of women marrying by the end of the child-bearing period which had been characteristic of eastern Europe. That was the first convergence. Changes in the sex ratio of this population in the marriageable ages contributed to these developments and explain why the increase in marriage propensity has been greater for women than for men. These changes in the relative numbers of marriageable men and women have been the result of a series of events: emigration, war losses, changes in mortality favouring the survival of males to adult ages, and changes in the flow of births. There are countries which are near to being exceptions to this convergence in that though marriage propensity is high (there are relatively few spinsters left at ages 45–49), age at marriage is still somewhat high, but these countries are few (Ireland, Italy, Japan, Portugal, Spain, Sweden).

This change in itself, by increasing the proportions of these populations which consisted of recently married couples and by increasing the proportion of women who had married younger and had longer reproductive lives, was bound to lead to a rise in the overall level of fertility and it did; but in addition there was an increase in fertility at almost all ages at marriage above the low levels that had been reached in many countries in the 1930s at the end of a long-term decline. This increase, occurring while increasing proportions of couples were using birth control, initiating that use at an earlier stage in married life and shifting from less to more effective techniques of contraception, must have reflected a rise in the desired family size. There were, however, other countries which, pre-war, had not reached their nadir in fertility, e.g. Netherlands, Italy, Greece, Yugoslavia, which have shown a further decline since the war.

There was another convergence. The change in marriage habits in the west, and the contrast between the falling fertility in southern and eastern Europe and the recovering fertility in the remaining developed countries, narrowed the regional differences in both birth rates and marital fertility rates. More recent changes have maintained the convergence. In east Europe fertility has largely continued to remain low. Elsewhere in the United States,

Canada, and Australia, as well as generally in western, central and southern Europe, the early post-war stabilization and rise has been succeeded by a new decline. Part of the decline is due to some lengthening of birth spacing (after the post-war shortening) but Glass's view is that the decline has proceeded beyond the extent that would be explained in this way – 'the run of the data suggests that ultimate family size is also likely to be affected, and quite dramatically in some communities where hitherto family size had remained fairly high'. Convergence between regions has been assisted by convergence *within* countries. The spread of industrialization and urbanization, and reduction in illiteracy, have reduced the scope of differences in fertility between parts of the same country.

There is one other convergence of great importance – a reduction in the dispersion of family size. In the post-war recovery and stabilization of fertility, the large family did not return. The frequency of childless and one-child marriages fell, and there was a rise in the proportions of two and three child marriages. Families have tended to become standardized in this new two-three norm determined by contemporary social and economic conditions in which couples have the capacity to fix and achieve targets in family size that take account of the wider range of social roles of women and the wider social and economic aspirations of couples.

But this convergence brings us to an important condition of modern fertility upon which Professor Glass does not comment – the extension of choice. Already, for a decade or two in Western society, marriage and motherhood has ceased to be the only career for women; there are now many roles open to them. Already for a decade or two a married woman has had the knowledge, the means and the freedom to choose whether or not to bear a child without having to sacrifice a healthy sexual life (admittedly there are still pronatalist pressures both economic and social, but the former have never been very effective and the latter are weakening – even the woman's magazine writers have eased their emphasis on 'fulfilment' of motherhood). Now there is the possibility of the maintenance of low average family size and of low population pressure (so necessary in growing world pressure on resources) while permitting a return to a wider range of family size. With those who do not want families, freed of the stigma of 'not doing their duty', 'fulfilling themselves', etc., and able to fill other roles, the way is open for those who really want large families (and who are likely to be highly motivated parents) to have them without disturbing the low average. A new dimension of choice.

How low is the average at the present time? Professor Glass does not and could not provide a complete answer to this question. He does, however, provide in an Appendix three important sets of figures. First is a table of gross reproduction rates (numbers of *female* live births per woman surviving throughout the reproductive period up to age fifty) at approximate periods of time from 1931–3 to 1961–6, and from 1964–73, for all the developed countries. In the 1970s many of the figures are only just above unity and in a few cases even a little below – representing a family size of little more than two. A second table provides generation fertility rates for selected countries (total fertility by age fifty) for successive generations of women. Again, per woman, the figure for women born in the late 1930s is down to low levels: Belgium and Denmark (2·4), England and Wales (2·2), Finland (2·1), Netherlands (2·2), Norway, Sweden (2·0), Switzerland (2·3), Italy, Czechoslovakia (2·1), Hungary (1·9), USSR, Canada (2·1), Japan (2·1). But the latter figures are accumulations of actual experience influenced by events in the not-so-recent past. The third set of figures is of duration-specific marital fertility rates (average number of live births for married women by date of marriage and by duration of marriage). The figures are necessarily incomplete for recent marriages – those married in 1965 and later can only have been observed for the first few years of marriage. But the figures confirm the trend to the small family norm. For recent marriages the England and Wales figures project to a completed family size of almost 2·2 – just above replacement.

Professor Glass devotes a major part of his lecture to a review of the factors that have brought about these changes. First is the acceptance of the fact of fertility control even in strongly pronatalist countries like France. Second is the increased availability of more effective techniques of contraception (accelerated by mounting concern about rapid population growth in developing countries which lent respectability to research into their techniques and their application). Third is the concern about world population growth which also produced an ecology front. Fourth are the many aspects of the growing emancipation of women – their wider entry into the labour market and their demand for alternative roles, their pressure for wider provision of birth control facilities, including abortion. Professor Glass provides much factual material to fill out the history.

What of the near future? Professor Glass does not expect any significant changes in marriage patterns. Within marriage, the prospects are considered under two headings, birth control and social and economic influences on the motivation in respect of desired family size. As to the former, a spread of more effective techniques is expected. As to the latter, Professor Glass considers that the extension of education, the movement toward skilled manual and white-collar jobs, greater employment opportunities for women, growing industrialization and urbanization in the less-developed countries, will lead to further convergence to low-fertility levels. But there may be fluctuations – the corollary of greater fertility control is greater responsiveness to economic changes. Professor Glass considers the possible impact of explicit government population policies whether, as in most cases, to reduce growth rates in the face of the mounting problem of world population pressure (in other developed countries as in Britain these policies are of a low profile character), or, as in some eastern European countries, faced with what they regard is an excessive fall in fertility, to encourage births, e.g. by restricted access to contraception or abortion. He regards such policies on the basis of past experience to be likely to have only marginal impact: '. . . even radical restrictions may fail to produce the desired results if there is powerful counter-motivation on the part of husbands and wives'. It comes back to the crucial issue of personal choice.

## The world distribution of the human polymorphic variants

The recent appearance of the second edition of Dr Mourant's *The Distribution of the Human Blood Groups* (1976) provides an opportunity to appraise the contribution to human genetics of knowledge of the spatial distribution of polymorphic variants. This was still a new field in 1954 when the first edition appeared. The book came at an opportune moment, and provided an excellent and highly useful compilation of the then known frequencies of human blood group genes from studies in a number of populations, a compilation, moreover, in which the data had been screened for quality and comparability, while its text illustrated the variety of specific anthropological problems to whose solution the data contributed.

Examination of the material in the tables and maps allowed the establishment of a number of fundamental generalizations. First, all human populations show genetic variation among their component individuals, and the variations in gene frequency among human populations are far from random; in system after system the frequencies of the genes cluster in a relatively restricted group of all possible frequencies. Secondly, in the spatial distribution of frequencies, general geographical patterns can be discerned. Many of the characters show strong clinal arrangements, though with some variation in frequency about the gradient itself. These gradients indicate that geographical distance between populations is an important determinant of their gene frequencies. Thirdly, major steps in the gradient occur from time to time, and where they do they usually separate one major continental group of man, or one clearly defined ethnic entity, from another. Sometimes these steps coincide with geographical barriers, for example the marked discontinuity of the Sahara separating Mediterranean frequencies in the Rhesus system from those of sub-Saharan Africa, or the break of slope in the MN system in the region of Wallace's Line separating Australasia and the eastern parts of Indonesia from the western end. Fourthly, the terminations of the gradients, i.e. where the maximum and minimum frequencies occur, differ from one character to another, so that for example the maximum frequency of the blood group B gene occurs in central Asia, that of gene C of the Rhesus system in the New Guinea region. Hence no one human population can be picked out as being conspicuously different from all others, and this in turn implies that all have advanced similar distances along the evolutionary pathway but in slightly different directions. Fifthly, some genes show a remarkably wide range of frequencies, for example the gene C of the Rhesus system, others (e.g. E of that system) much less variation.

How great have been the advances in such descriptive population genetics of the last twenty years! This is shown by the new edition of the book. There is the sheer mass of new data collected; instead of 82 pages of tables as in the first edition there are now 660. Partly this is due to the new blood group systems that have been discovered, such as the Diego, Yt, Auberger, Dombrock and others, the majority of which have already been investigated in several populations. Then there are the numerous new subdivisions due to discovery of further alleles in existing systems, which have necessitated the subdivision of tables according to whether or not a particular antigen was tested for. Thirdly, there is the enormous dividend from the development of gel electrophoresis, the numerous serum protein and red cell enzyme systems, many of which show considerable population variation in frequency of the alleles responsible for their variants. The tables would have been still more numerous had there not been deliberately excluded the complexities of the immunoglobulin groups, particularly the Gm and Inv antigens, and of the more recently discovered loci of the histocompatibility system.

Undoubtedly much of this advance was directly stimulated by the publication of the first edition, which has kept busy a generation of investigators. Undoubtedly also the new edition, which brings together in such a convenient form so much scattered data, will provide ample material for a new generation to investigate a variety of problems of genetic evolution – population affinities, derivations, genetic distances, environmental associations, as well as specific anthropological problems – by examining these data alone or correlating them with other lines of evidence. It picks out·quite clearly the areas of the world's surface in which the genetic constitution of the populations is still too little known, and thus will give some guidance to future field expeditions. The text follows the same pattern as previously, an introduction to the different systems, followed by a chapter by chapter discussion of regional populations of a primarily factual nature, followed by an attempt at a synthesis. But Mourant now includes his exciting speculations in the regional texts, and restricts his final discussion to pointing out how little is known of the biological meaning of blood group and protein variation with suggestions for further research.

This is indeed a weighty volume. At 6 lb instead of the previous $1\frac{3}{4}$ lb, it is massive, and academically it is monumental. It represents a lifetime's work of one man, aided by his loyal associates, and the stimulation and help that he has so unselfishly given to so many others. The second edition validates the first, for the original generalizations that could be drawn still stand, though in rather more refined form. But the achievement goes much further. Few other studies have made so noted a contribution to our understanding of the nature and meaning of human genetic variation.

## Notes

1 In all these calculations childlessness in married couples is taken into account (i.e. family size *zero* is included in the average), but not differences in the proportion of women in different social groups who marry (and who appear in the denominator of the average). This proportion is lower in the lowest social class than in the highest, but not, in the classes used here, substantially so – certainly not to such an extent as to counterbalance, in terms of reproduction of the group, the much larger differential in family size. We have been observing (1951) a family size in the highest social class which is two-thirds that of the lowest, at a time when, at most, the marriage proportion is only of the order of 10 per cent higher. Differential marriage incidence is not an important factor in this particular context (a short note on this point appeared in *The Eugenics Review* (1966), **58**, 49–50).

# PART THREE

# *IQ and Social Stratification*

# 9 The inheritance of inequalities: some biological, demographic, social and economic factors

*J. E. Meade*

(From *Proceedings of the British Academy*, vol. 59, 1973.)

In a society in which there were no governmental interferences with the operation of the competitive market and no other artificial impediments to competition or mobility, persons who were similarly endowed would tend to receive the same incomes.

But if individual citizens are not equally endowed, then personal incomes may continue to be unequal even in a fully competitive, *laissez-faire* society with unrestricted mobility. The man with little skill and ability will not necessarily be able to undercut the man with great skill and ability, even though the earnings of the latter greatly exceed those of the former. The man with much property will have a higher income from property than the man with little property even though the rate of return on all properties were the same.

In this paper I wish to isolate for examination some of the factors which would cause citizens to be unequally endowed and thus to receive unequal incomes even in a competitive, *laissez-faire* society with unrestricted mobility. For this purpose I shall proceed for the most part as if there were free competition, unimpeded mobility, and no governmental interference in the economy; and, on these assumptions, I shall inquire what influences one would expect still to remain to cause inequalities in personal endowments of income-earning factors of production. I am not thereby intending to assert that the actually existing structure of inequalities can be explained without allowing for the influence of such factors as customary ideas about fairness which may cause rigidities in pay differentials, or impediments to movements from a low-paid to a high-paid occupation, due, for example, to trade union or similar restrictions on the entry of outsiders into a protected occupation, or governmental tax policies and similar measures, many of which are expressly designed to affect the distribution of incomes and properties. I am merely engaged in one preliminary exercise of abstraction which may help to highlight certain important influences which must be brought into any final calculation.

A citizen in a *laissez-faire* competitive society would receive certain endowments from his parents which would help to determine the amount of income which he could earn and property which he could accumulate during his own lifetime. This in turn would affect the endowments which he could hand on to his children.

The endowments with which we will be concerned may be enumerated under four heads.

First, a citizen will be endowed with a certain genetic make-up. There is some genetic component in intelligence which may affect earning capacity. But it would be a mistake to forget other characteristics which probably have some genetic component and which may well exert a greater influence on earning capacity. Quite apart from straightforward bodily strength and health, there may be other relevant physical differences which have some genetic component; there may, for example, be some genetic influences affecting the vocal cords of Mr Fischer-Dieskau and Miss Janet Baker which help to explain their ability to earn income. There may also be genetic components in the determination of certain qualities of character which have an income-earning potential, though it by no means follows that all of these are desirable in themselves. Thus, a certain streak of ruthlessness and aggression may be helpful to the accumulation of wealth without being in any basic ethical or aesthetic sense good or desirable qualities in and for themselves.

Second, a citizen may inherit a certain amount of income-earning property of one kind or another from his parents.

Third, a citizen will have received as a child a certain education and training. In a strictly *laissez-faire* competitive society this education and training will have been provided and financed privately by his parents, though this is, of course, one of the fields in which my neglect of governmental interventions and policies is especially significant.

Fourth, these are the rather less tangible advantages or

disadvantages which accrue to a citizen through the social contacts which he makes with other persons, these social contacts being much affected by the social background into which he was born.

These two last elements of endowment – namely, education and social contacts – must in my scheme of things cover a very wide range of social phenomena. Education obviously covers an individual's formal education and training at school, university, or similar institution. Social contacts obviously cover a citizen's range of acquaintances who through their particular brand of the old-boy network can or cannot get him a good job or provide him with a favourable investment opportunity. But there are many other factors to be taken into account which in my limited scheme must be put into either the one or the other of the very general categories of 'education' and 'social contacts'.

I personally think of the category of education as covering practically all of the environmental influences which affect the development of an individual's knowledge, character, and motivation. He will thus receive much of his so-defined education directly from his parents as they bring him up in a certain way and from the acquaintances he makes – to say nothing of the education which a husband receives from his wife, and which a wife receives from her husband.

If education is defined in this very broad way, then social contacts must be narrowly defined and are reduced to little more than a catalogue of the sort of friends, acquaintances, neighbours, and colleagues with whom an individual spends his days.

A citizen is thus fortunate or unfortunate according as he starts out in life with a helpful or unhelpful endowment of genes, inherited property, education, and social contacts. But in addition to these initial structural elements of good or bad fortune which are determined by his family background, a citizen will also encounter many elements of good or bad luck in the course of his career. To take but one example, two men with the same inborn ability and the same initial advantages of education, property, and social contacts may end up with very different incomes and properties, simply because they embarked on careers in different lines of economic activity, one of which prospered and the other of which declined. And yet at the time of choice the prospects of the two activities may have seemed very similar to both of them and it may have been a matter of almost random chance which determined the choice of career. In what follows I shall use the term 'fortune' to describe the basic structural endowments of genes, property, education, and social contacts, and the word 'luck' to describe the many chances in life which determine the actual outcome within these structures of basic endowments. One cannot, of course, draw any hard and fast line between elements of fortune and elements of luck as I have tried to describe them; they are both mixtures of recognizable laws of cause and effect and of strokes of pure chance; but the nature of society – or should I say of the social studies? – is such that it seems to me useful to think in terms of some such broad distinction.

Social scientists examine the general genetic, demographic, social, and economic structure of society. They consider the characteristics of, and the factors affecting, various groups: income groups, property groups, IQ groups, social classes, age and sex groupings of the population, occupational classes, classes of educational attainment, and the like. $A$ may be born into one set of groupings and $B$ into another. When the souls of little $A$ and little $B$ were lining up in heaven to be sent forth on their sojourn in this wicked world, did they toss up as to which soul should occupy which niche in the social structure which they were joining? I do not know. But I shall refer to the structured endowments which $A$ and $B$ receive in society by joining whatever group they do join as their good or bad fortune.

However, different people within the same niche in the structure of society may fare very differently in the course of their lives. It is the causes of these divergencies in the fates of two persons within the same fortunate or unfortunate structural niche which I shall call factors of luck. This is not to assert that these factors are in any fundamental sense less subject to laws of cause and effect than are the factors of fortune. My category of luck certainly contains all those causes of inequality which are not explained by the structured influences of what I have called fortune; and there may well be disciplines other than present-day economics and sociology which would help to explain why two persons with the same structured fortune fare differently in the outcome.

The basic structural endowments of good or bad fortune are handed down from parent to child; but the child as he grows up moulds and modifies the basic endowments which he received as a child from his mother and father, before he amalgamates them with those of his wife and passes this package of *modified* and *mixed* endowments of fortune on to his own children. I will start first with a consideration of the way in which an individual's initial endowments may be *modified* as he grows up; and I will turn later to the implications of the fact that he *mixes* these *modified* endowments with the already *modified* endowments which his wife received from her parents before the two of them hand on this modified mixture to the next generation. (I have put this example in terms of a boy only because the English language does not possess a pronoun which covers both male and female. Solely for this reason, in what follows I shall analyse in terms of the male sex much that applies equally to the female sex.)

Let us then consider how a citizen's passage through life may affect the elements of basic structural fortune with which he was initially endowed. This is illustrated in Figure 9.1, in which I consider the way in which a particular citizen (let us call him Tom Jones) starting out as little Tommy receives his basic endowments from his home background, proceeds through life, and at length as poor old Tom, or Thomas Jones Esquire, or maybe even Sir Thomas Jones, GCB, himself contributes to a home background transmitting endowments to his children.

Tom Jones then starts in a home background ($H_1$) which is built up by his father and mother. We are concerned with his parents solely as instruments affecting his basic endowments of good or bad fortune; and in this sense his father and mother are themselves simply bundles of factors which will affect their ability to provide Tom Jones with his initial endowments of fortune. The parents' relevant factors I assume to be the mother's and the father's genes ($G_m$, and $G_f$) (line 1), education ($E_m$, $E_f$) (line 2), and social contracts ($C_m$, and $C_f$) (line 4), and their joint income ($Y_j$) and property ($K_j$) (lines 3 and 5). These together constitute the home background which provides Tom Jones with an endowment of genes ($G$), education ($E$), social contacts ($C$), and property ($K$). Thus in the diagram we look upon the home background as a GEYCK which produces a GECK for each child.

One must not, however, regard this endowment of Tom Jones by his parents as a once-for-all affair which occurs instantaneously at his birth. It is a continuing process; and this introduces two interacting dynamic factors. In the first place, Tom Jones will be susceptible to different endowments at different stages of his life: to his parents' genes once-for-all on his conception, to the qualities of his mother's care as an infant, to his parents' friends at

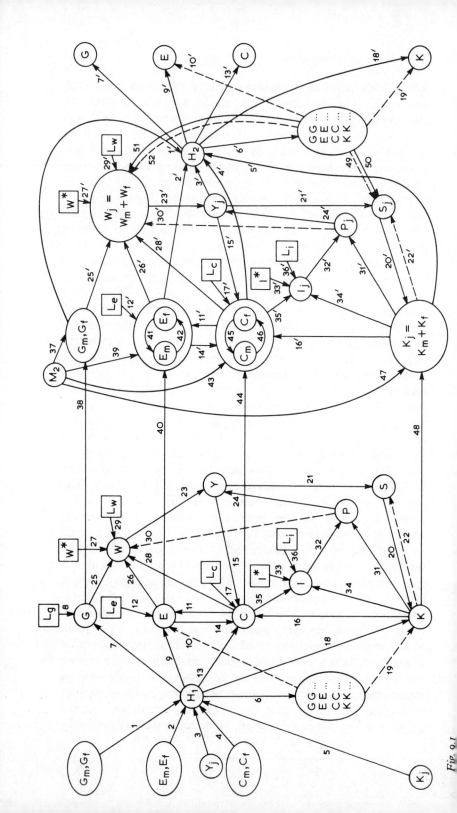

Fig. 9.1

a later stage, to his inheritance of property on his parents' deaths, and so on. Second, his parents' own education, income, social contacts and property will be developing during their years as home-builders and parents, so that what they have to give as well as what Tom Jones is ready to receive will be changing over time.

I shall at first neglect the influences affecting what Tom Jones's parents have to contribute and shall take the nature and development of his parents' genes, education, income, social contacts, and property as given. I shall return to this set of problems when I close the cycle and come to regard Tom Jones himself as a parent. I will then consider his development as a provider of endowments for his children. For the time being I wish to consider him solely as a recipient of a given developing flow of basic endowments from his parents, which he himself then develops further.

To return to the diagram, Tom Jones's parents may produce brothers and sisters for him, and these are represented by the little GECKs which proceed from Tom Jones's home background $H_1$ (line 6). But the main purpose of my diagram is to put the individual Tom Jones under the microscope.

From his two parents Tom Jones receives his genetic endowment ($G$) (line 7). But while his genetic make-up is basically conditioned by that of his parents there is also an element of luck ($L_g$) (line 8). Two children of the same parents will not receive identical genetic endowments unless they are identical twins. Tom Jones can draw his genes only from those offered by his parents; but he may have good or bad luck in his draw from the parental stock.

Tom Jones will receive an education ($E$). In the absence of governmental intervention not only much of his early upbringing but also his formal education and training will be provided for him by his parents (line 9). However, the greater the number of Tom Jones's brothers and sisters, the less his parents may be able to afford out of their given time, income and property to invest in Tom Jones's individual education (line 10). In addition to his home background and formal education, much of what I have broadly defined as his education will be continued during his own career by his social contacts, that is to say, by the sort of friends and colleagues with whom he associates (line 11). But in all this there is a considerable admixture of luck ($L_e$) (line 12). To take only one example, his parents may make most

carefully calculated decisions about the amount of money which
they will invest in his education and about the educational
institutions to which they will entrust him. But the outcome
may be greatly affected in ways which it may be impossible to
foresee by luck – as, for example, whether a particular teacher
happened to fire young Tommy's imagination and interest in a
particular subject or activity.

Tom Jones will inherit certain social contacts ($C$) from his
parents (line 13), since the social environment of his home
background will greatly affect his choice of friends and acquaintances. But as he grows up his social contacts will develop and
will depend upon the way in which his own career develops. An
important factor will be the social contacts which he makes at
school or other educational institutions (line 14). Thereafter, the
further development of his social contacts is likely to be affected
by his material success in life. If he manages to earn a high
income ($Y$) or to acquire much property ($K$), the fact that he is
a man of riches will enable him to make contacts with people
who will be useful to him in his career (lines 15 and 16). Finally,
of course, there is an important element of luck ($L_c$) in the
people he meets and the friends he makes (line 17).

From his parents Tom Jones may also receive property ($K$)
(line 18). But once again the greater the number of Tom Jones's
competing brothers and sisters, the smaller will be his own share
of the family property (line 19). As time passes he may
supplement this property from his own savings ($S$) (line 20). The
level of these savings will be affected by many considerations; and
in the diagram I have introduced only two of the most important.

In the first place, the higher his income is the greater will be
Tom Jones's ability to save (line 21).

But, in the second place, the greater the property which he has
already acquired (perhaps by inheritance), the smaller will be his
need to save, since there will be less need to abstain from present
consumption to acquire a property to support him in his old age or
to give him security against adversity. This fact that the higher his
property ($K$) the lower will be his savings ($S$) is represented by
the broken line 22.

It remains to consider the factors determining the level of Tom
Jones's income as he passes through life. His income ($Y$) is simply
the sum of his earnings or income from work ($W$) and of his
income from property ($P$) (lines 23 and 24).

His earnings will be affected by many factors. First of all there is his capacity to earn which will be affected both by his genetic endowment and by his educational endowment (lines 25 and 26). But, given his ability, his actual earnings will depend upon the structure of wage rates that exist in the market for different kinds of ability ($W^*$) (line 27). Earnings, however, are not determined exclusively by a given market wage rate for a given ability. There is an element of fortune in that good social contacts may enable a man to make a more rewarding choice of job (line 28); and there is also an element of luck ($L_w$) in determining whether Tom Jones will be successful in his choice of occupation or in the development of his particular job (line 29).

There are other important influences in the real world which I am neglecting as a result of my assumption of free competition – influences such as trade union or similar restrictions on entry into protected occupations or customary differentials in pay which interfere with market forces. But there is one further consideration which I cannot neglect in my competitive economy.

Tom Jones's earnings will depend in part upon the amount of effort which he chooses to put into the business of earning a high income. This is influenced by many factors; but among these we may suppose that the higher is Tom Jones's income from property ($P$), the lower – other things equal – will be the effort which he puts into earning an income from work (line 30). Indeed, if he has a sufficiently high income from property he may not bother to earn any additional income at all.

At this point I must digress to ask myself whether my diagram covers the undoubted fact that Tom Jones's own moral character and motivation will affect how hard he will work and what steps he will take to get on. Do not some people get on – and deserve to get on – because they try hard, and others fail to make good – and deserve to fail – because they make little or no effort to help themselves? We are immediately faced by the riddle of free will. Do not a man's genetic and environmental endowments, together with some elements of pure luck, for which he can in no way be held responsible, determine his moral character and motivation as well as his ability? If so, it can all be comprehended in lines 25, 26, 28, and 29 of my diagram. But it would then seem meaningless to assert that Tom Jones was in any way a free agent in deciding whether to deserve success or failure. But if one does believe in some measure of free choice and personal responsibility for success or failure – and I cannot help doing so – there is something vital – but I do not quite know what – missing from my diagram. This is one of those many difficulties which I learned from Professor Sir Dennis Robertson to look squarely in the face and pass on.

Let us turn now from Tom Jones's earned income to his income from property ($P$). This is simply his property ($K$) multiplied by the average yield or rate of interest on it ($I$) (lines 31 and 32). The yield on property will be basically determined by the structure of the ruling market yields on various types of property ($I^*$) (line 33). But the actual yield obtained may well be affected by Tom Jones's investment opportunities. Thus the yield on property is likely to be higher for a man with much property to invest (line 34) and for a man with the right social contacts (line 35). A man with a large property can afford to take more risks in his investments, and the cost of advice from stockbrokers and of other investment services can be spread over a larger capital fund. For these reasons a large property normally obtains a higher yield than a small property. Moreover, a wealthy man is more likely to have those social contacts which will enable him to be better informed about the chances of profitable investment. Finally, let me point out, in case any of you have not operated on the stock exchange, there will be an element of luck in Tom Jones's choice of investments for his property ($L_i$) (line 36).

Tom Jones grows up into mature manhood with a certain make-up of genes, education, income, property and social contacts, these elements of his make-up being, as we have seen, partly inherited from his original home background and partly made up by his own social and economic development. He is now ready to marry a wife and to become a father; and together these two bundles of genes, education, income, social contacts, and property, having joined together in holy matrimony, are ready to make up a second-generation home background for the next generation of children.

I will turn to their married life in a moment. Let me pause for a little to comment on my account of Tom Jones's bachelor life.

A very marked feature of the simple model which I have presented in the diagram so far is the amount of positive feedback which it contains – that is to say, of self-reinforcing influences which help to sustain the good fortune of the fortunate and the bad fortune of the unfortunate.

Let me give two examples.

The first concerns job opportunities. A man who for any reason starts with a high income may be able to make appropriate social contacts which enable him to find exceptionally repaying jobs which will in turn help to raise his income still further (lines 15, 28, 23).

My second example concerns the accumulation of property. A man who for any reason of good fortune has a high income can save much and accumulate a large property (lines 21 and 20). But with a large property he has a high income from property (line 31) and thus a still higher income (line 24). Nor is that the end of the matter; with a high property he can probably get a high yield on his property, partly because a large property can be more cheaply and effectively managed than a small property (line 34) and partly because a man of wealth will be better able to make the sort of social contacts which will enable him to invest his property profitably (lines 16 and 35). Thus the yield on his property, as well as his property itself, will be raised simply because his initial fortune was good.

This particular set of positive-feedback relationships probably helps to explain one of the very pronounced phenomena in our type of society – namely, the very much greater degree of inequality in the distribution of the ownership of property and of income from property than in the distribution of earned income. An individual with a high income is able to save a higher proportion of his income than can an individual with a low income. A man with high earnings will thus accumulate a property which is high relative to his already high earnings. If, having a high property, he then gets an especially high yield on property, his income from property will become large relative to his property which will become large relative to his already high earnings. Conversely, for the citizen with low earnings, his income from property will be low relative to his property which will be low relative to his already low earnings. The discrepancy between high and low property incomes will be much greater than the discrepancy between high and low earnings; and, to anticipate my analysis, these discrepancies are likely to be perpetuated from one generation to another through the inheritance of properties and earning capacities.

My diagram has many positive-feedback loops. It contains through the broken lines 22 and 30 only two examples of negative feedbacks, of influences, that is to say, which damp down rather than multiply the results of initial good or bad fortune. Thus it is probable that the higher a man's property is, the smaller is his incentive to cut back his present consumption in order to save and accumulate; and this factor damps down the way in which large properties tend to lead to still larger properties (line 22). In a rather similar manner the existence of a high income from property reduces the need for income from work and may thus damp down the incentive to earn more (line 30); and this factor may reduce the positive, reinforcing effects which we have just examined, whereby high incomes lead to still higher incomes.

However, my assumption of *laissez-faire* has forbidden me to display on my diagram some fundamental elements of negative feedback which may be at work in the real world through governmental interventions. Progressive taxation, the provision of free education and medicine, and the payment of social security benefits or other supplements to the incomes of those who are less well off, in so far as they are effective in redistributing income from the rich to the poor, are outstanding examples of such negative feedbacks. In such circumstances a rise in a man's gross income and wealth (before governmental adjustment) causes a less than proportionate increase in his net income and wealth (after governmental adjustment); and this diminishes the multiplier whereby initial good fortune feeds upon itself and magnifies the final outcome.

But there remain in society very strong elements of positive feedback which I have illustrated in my diagram. Two results follow from this.

First, there is the obvious point that there are some apparently powerful built-in tendencies for the rich to sustain their riches and the poor their poverty which one would expect to help in explaining the persistent continuation of the large inequalities in income and wealth which we actually observe in society.

A second major result may be expected from the intertwining of the many positive feedback loops in my diagram, namely that the various endowments passed from parent to child are likely to become highly correlated with each other. Thus if Tom Jones is born with a set of useful genes which help him to earn a high income this will enable him to make useful social contacts and to accumulate a sizeable property. Thus as a father he is likely to be a bundle not only of useful genes, but also of a useful income, a useful property, and useful social contacts. There will be a strong tendency in society for good or bad fortune to be handed on to the next generation in associated parcels of genes, income, property, and social contacts.

This tendency for the useful endowments of various kinds to become associated with each other will be further strengthened when we allow for the mixture of Tom Jones's endowments with those of his wife. Tom Jones marries Mary Smith. Tom Jones may be fortunately endowed with an educational and genetic make-up which turns him into an able, enterprising, perhaps ruthless, but anyhow successful businessman. Mary Smith may be fortunate in being the heiress to much property and endowed with the best social contacts. If in society there is a tendency for the fortunate to marry the fortunate and for the unfortunate to marry the unfortunate, whatever may be the primary cause of their good or bad fortune, then there will be a tendency for Tom Jones's useful genes and education to be joined with Mary Smith's useful property and social contacts. The various elements of basic endowments will become more highly correlated with each other.

But I am anticipating the next stage of my analysis. If the new biologists had already succeeded in getting rid of sex as a method of human reproduction, I would have little to add to the analysis presented in the first half of my diagram. If Tom Jones by some process of cloning could by himself produce a little son with an exact replication of his own genes, we could explain most of the factors affecting the development of inequalities of income and wealth as between various families by concentrating solely on the influences which I have discussed so far. Tom Jones would receive his endowments of genes, education, social contacts and property from his father. He would hand these same genes on to his son; subject to all sorts of luck, he would develop his property and social contacts in the way which we have examined and, in the light of this development of his fortune, he would pass on an education, social contacts and property to his sons. The situation could be much affected by the number of sons which he decided to clone – I will return to that subject in due course – but apart from that there would be little to add to the analysis.

But the fact that he has to marry Mary Smith and mix his genes, income, social contacts and property with hers before they jointly endow their sons and daughters introduces many basic modifications into the analysis. We will start the analysis of these problems by assuming that Tom Jones has chosen, or been chosen

by, a particular Mary Smith with her own particular bundle of
genes, education, social contacts and property as they exist at the
time of her marriage. I will discuss later the very important
question of what it was that brought Tom and Mary together.
For the moment I am interested in the implications of their joint
family life.

A family is more than a number of individuals. In the first half
of my diagram we watched the development of Tom Jones as an
individual bachelor. In the second half of the diagram we watch
Tom and Mary Jones's family developing as a joint concern.

I represent Mary Smith as $M_2$, namely as a second-generation
mother. For our purpose she is simply a bundle of factors relevant
for the joint building of a second-generation home background
($H_2$) for the endowment of the second generation of children. She
brings into the marriage her genes, education, social contacts and
property, the nature of which depend upon what endowments she
has received from her parents and the way in which she has
developed them during her spinsterhood.

Thus Tom and Mary together provide a pool of mother's genes
$G_m$) and father's genes ($G_f$) for use by the family (lines 37 and
38). They provide mother's education ($E_m$) and father's education
($E_f$) to form part of the family background (lines 39 and 40).
Their educations in my broad sense of the term continue during
their married life; and this is partly due to the fact that they
educate each other (lines 41 and 42). They provide the mother's
and the father's social contacts for use by the family ($C_m$ and $C_f$)
(lines 43 and 44); and Tom's contacts enlarge Mary's contacts
and vice versa (lines 45 and 46). They both bring some property
into the family ($K_m$ and $K_f$) (lines 47 and 48); and I am assuming
that they form a close-knit family in which the two properties are
for practical purposes merged into a single joint family property
($K_m + K_f = K_j$) with a corresponding joint family income from
property ($P_j$) derived from the yield on the joint family property
($I_j$). Similarly I assume also that Tom and Mary merge their
individual earnings into a joint family income from work
($W_m + W_f = W_j$). Thus there is a joint family income ($Y_j$) from
which joint family savings ($S_j$) are made.

The main relationships within this family are now exactly
similar to those in the first half of my diagram. I will not go into a
tedious repetition of the strokes of luck which Tom and Mary may
find in their further education, their social contacts, their

investments or their jobs, nor with the way in which the various elements in their family structure feed back upon each other. The relevant lines in the second half of my diagram correspond exactly to the same relevant lines in the first half of my diagram. (In the diagram I have made this clear by numbering the relevant lines on the right-hand half of the diagram with the same numbers as the corresponding lines on the left-hand half of the diagram. Thus line 1 on the left-hand is numbered $1'$ on the right-hand half; and similarly for the other numbers.)

All that part of the second half of the diagram is a mere application to the joint family of the relationships considered at some length in the case of Tom Jones's bachelor life. But there is now an important additional consideration to be introduced.

At the far right of the diagram we have a new home background $(H_2)$ made up of Tom's and Mary's genes, education, income, social contacts and property as these develop during their married life (lines $1'$, $2'$, $3'$, $4'$, $5'$); and these provide endowments of genes, education, social contacts and property as Tom's and Mary's little GECKs are born and grow up. If we want now to consider the life cycle of one of these in particular (for example, the life cycle of Tom's and Mary's son Richard), we start from the large GECK at the far right of the diagram, which shows Richard Jones endowed with genes, education, social contacts and property from his home background $(H_2)$ (lines $7'$, $9'$, $13'$ and $18'$), but competing for education and property (lines $10'$ and $19'$) with his brothers and sisters represented by the other little GECKs proceeding from the same home background (line $6'$). We have in fact cycled back to the extreme left-hand end of the diagram, but for generation 2 instead of generation 1.

But the size of Tom's and Mary's family will feed back into their own development as parents. The larger their family, the greater their financial responsibilities for feeding, clothing, housing, entertaining and educating their children. The greater these responsibilities, the more difficult will it be for them to save and accumulate property. The broken line 49 represents the fact that the larger the number of their children is the more difficult will it be for Tom and Mary to accumulate property during their married life. It is probable that they will in fact accumulate a smaller property. But this is not absolutely certain since, while their ability to save will be less, their motivation to save may be greater, since the larger the family the more they must accumulate in order to be able to give each child an inheritance of any given absolute size. This increased motivation for saving is shown by the continuous line 50.

The size of their family will also affect their earnings. A large family may make it more difficult for Mary to go out to work and earn an income. On the other hand it will increase the need for income and may increase the parents' motivation to seek as high an income from work as they can manage to earn. The net result is uncertain and I have represented this by a solid line 51 which represents the number of children as increasing the motivation to earn income and a broken line 52 as reducing the mother's opportunities to earn income.

My diagram is complicated enough; but even so it is a great simplification of reality. There are causal relationships which I have omitted from my diagram. Thus I have not allowed for the fact that a man's genetic and educational background may affect his ability and his effort in investing his property so as to obtain the highest possible rate of return on it; nor have I allowed for the fact that a man may during his career invest resources in his own further education and training, his ability to do so depending upon the level of his income and property. It would be easy to add the arrowed lines to my diagrams which would represent these further positive feedbacks; I have refrained from doing so simply in order to keep the picture clear.

Moreover there are certain other very important relationships which are perhaps implied in my diagrams but which are not very clearly represented in them and which I have not discussed. Thus my diagram fails to bring out the fact that the endowments which parents can give to their children may compete with each other. The more money a parent invests in a child's formal education ($E$) the less he may be able to leave to him in the form of other income-earning property ($K$). Moreover, parents who apply their minds to the direct care, education and amusement of their children at home may have less time and energy left for making money to leave to them.

Above all I have not discussed what determines the number of children which a set of parents will produce. It may well be that the structured genetic, educational, social and economic characteristics of the parents do influence the size of their families, some types of family having on the average a larger number of children than others. But there would almost certainly be important dispersions around these characteristic averages, the representation of which would need the introduction of yet another 'luck' factor. I shall have something to say later about the important effects of differential fertility between different types of family; but I have nothing to say about the *causes* of differential fertility. This is probably the most important omission from the general scheme of relationships which I am trying to put before you.

Finally, there is another very closely related demographic consideration. My diagram is based on the assumption of the permanent monogamous family in which Tom has children only by Mary and Mary has children only by Tom. This is still the basic pattern in our society, though the bonds of marriage are looser than they used to be. In a society in which human breeding pairs were frequently reshuffled the picture would be very different. In particular I would need to modify substantially what I am about to say on the mating patterns of husband and wife.

But let me return to my model with all its admitted deficiencies. I have now discussed how Tom as a bachelor and how Tom and Mary as a married couple develop the endowments which they received from their parents, mingle them into joint family endowments, and hand them on in turn to their children. Let me next turn to the important question of the factors which caused Tom and Mary to choose each other as mates in the first place.

I have already argued that there are strong forces at work in society causing the basic components of good or bad fortune – genes, property, and social contacts – to become highly correlated with each other; and I shall start my analysis of this question by talking of the fortune of a man or woman as if there were some single index of the amount of genetic–property–social-contact 'fortune' which a man or a woman possessed at the time of his or her marriage.

The fact that Tom Jones mingles his fortune with that of Mary Smith before he transmits endowments to the next generation will tend to limit the degree of inequalities in family backgrounds and endowments which would otherwise develop.

Let us imagine all the eligible bachelors drawn up in a strictly descending order of their fortunes and all the eligible spinsters similarly drawn up in a strictly descending order of their fortunes. We may say that there is perfect assortative mating if the most fortunate bachelor married the most fortunate spinster, the second most fortunate bachelor the second most fortunate spinster, and so on down the two lists.

In this case there would be no averaging of fortunes as the

generations succeeded each other. But consider, simply as an intellectual exercise, what might be called perfect anti-assortative mating. Suppose that the most fortunate bachelor married the most unfortunate spinster, the second most fortunate bachelor the second most unfortunate spinster, and so on down the bachelors' list and up the spinsters' list. The net result would be a tendency for the complete averaging of family fortunes in one generation, each family ending up with the same joint fortune (on the assumption that the fortunes of the bachelors and of the spinsters were symmetrically distributed).

Completely random mating may be defined as the case in which each pair of bachelor and spinster were drawn at random from the two lists.

In fact mating is somewhere between the completely random and the perfectly assortative. A bachelor at a given position in the bachelor's pecking order will not inevitably marry the spinster at the corresponding position in the spinster's pecking order; but the choice is not purely random; the nearer any given bachelor and any given spinster are to the same position in their two pecking orders the more likely they are to choose each other as mates.

But so long as mating is not perfectly assortative there is some averaging and equalizing tendency at work. If Tom's and Mary's fortunes do not correspond, then the joint family's fortune will be an average of whichever is the greater fortune and whichever is the lesser fortune. This is an equalizing tendency; and if this were the whole of the story, inequalities would progressively disappear as the generations succeeded each other. For as long as differences of fortune persisted there would be a force at work taking two different fortunes, joining them together, and averaging them. This force is known as the regression towards the mean. Exceptionally large fortunes would tend to be averaged with lower fortunes, and exceptionally low fortunes with higher fortunes. Fortunes would regress towards the average of fortunes.

If this regression towards the mean were the whole of the story we would expect to find society continually moving towards a more and more equal distribution of endowments. But there is another set of forces at work tending all the time to reintroduce inequalities, forces which we may call the forces of dispersion around the average. These forces are expressed in all the elements of luck to which I have drawn attention in my diagram – genetic luck ($L_g$), luck in education ($L_e$), luck in social contacts ($L_c$), luck in investments ($L_i$), and luck in one's work ($L_w$). If the genetic factors in ability were purely additive, then children would be likely on the average to inherit purely genetic factors for ability which were the average of their parents' genetic factors. But this is only an average. Unless they are identical twins, they will differ. Some will be lucky and some unlucky in the draw from their parents' pool of genes; and thus inequalities between the most and the least able in the family will be re-established. Moreover, in their careers some will strike lucky in education, social contacts, investments and jobs and will go uphill, while others will go downhill.

The ultimate self-perpetuating degree of inequality in the distribution of fortunes can thus be seen as depending upon the interaction of three forces. The less assortative mating is, the greater will be the regression towards the mean, and thus the smaller the ultimate degree of inequality. But elements of random luck in genetic make-up, and in social and economic fortune cause a dispersion about the average; and the more marked these elements are, the greater will be the ultimate degree of inequality in society. Finally, the more marked are the positive feedbacks and the less marked the negative feedbacks in my diagram of structured developments of endowments, the greater the ultimate degree of inequalities.

So far I have spoken in terms of a composite single index of fortune. But for many purposes it is necessary to break it down into its components. Consider the effects of changes in social habits which modify previously rigid social barriers. Suppose that members of different social classes begin to meet more frequently in clubs, sports, and other institutions.

Such changes would almost certainly make mating less assortative in terms of property and social contacts. The child of propertied parents with useful social contacts would be more likely than before to meet the child of propertyless parents with less useful social contacts.

But as far as ability to earn is concerned, whether this be due to genetic or environmental luck, the change might lead to greater assortative mating. In particular the introduction of a system of higher education which was less structured according to social class would tend to bring boys and girls together according to their intellectual ability. This would be particularly true of a university system which ceased to be a finishing school for the sons of gentlefolk and started to provide an education for the able sons and daughters of all classes. Only the able children of gentlefolk would get to the university where, for the first time, they would meet the selected able children of the working class – and this just at that impressionable age when it has been known for young men and young women to become fond of each other.

It would be tempting to conclude from this that such social changes might lead to a more equal distribution of property (as mating was less assortative according to property ownership) but a less equal distribution of earnings (as mating was more assortative according to those endowments which led to intellectual ability). But this overlooks the interconnections between the various endowments. High earnings lead to high incomes which enable large properties to be accumulated. It is possible, though not certain, that in the end the more unequal distribution of earning power leading to a more unequal chance of accumulating property would have so potent an effect in increasing inequalities in the ownership of properties that it would outweigh the equalizing effects on property of less assortative mating according to property ownership. The easier rise of the meritocratic élite and descent of the aristocratic dud might in the end increase the concentration of property as well as of income at the upper end, unless, of course, offset by governmental measures for the redistribution of income and wealth.

I turn to a second reason why we must distinguish between the different elements of good fortune. Until the new biologists have made further advances in their art, it will remain impossible for Tom and Mary Jones to control the genes which they pass on to their children. They cannot decide that little Richard shall inherit all the good genes and little Jane all the bad genes; little Richard and little Jane must both take part in the same lucky dip. But Tom and Mary Jones can decide that little Richard shall inherit all the family property while little Mary shall have none of it; and the laws and customs which regulate the inheritance of property can have a very important effect upon the ultimate degree of inequality in society.

One can illustrate this by means of the following artificial, but nevertheless suggestive exercise (based on Atkinson 1972, p. 63). Imagine a society in which there is no capital accumulation but a constant stock of property which passes by inheritance from generation to generation. Suppose this property to be shared initially in equal parcels among a privileged 5 per cent of the families. Suppose each set of parents in the community to produce the same number of children, equally divided in each family between boys and girls. Suppose every boy and girl to survive and to get married and to have in turn the same number of boys and girls as did their parents. If each family produces one son and one daughter, then the population will be constant. If each family produces two sons and two daughters, the population will grow, doubling in each generation.

We wish to watch the distribution of property as the generations succeed each other. Table 9.1 illustrates the way in which the combination of the degree of assortative mating according to property ownership, the growth rate of the population, and the laws and customs affecting the inheritance of property will combine to affect the outcome.

Table 9.1

| | Percentage of population owning property | | | |
|---|---|---|---|---|
| | Perfect assortative mating | | Completely random mating | |
| Properties left to: | Stationary population | Growing population | Stationary population | Growing population |
| 1 First son (*or* first daughter) | Percentage constant | Percentage falls (absolute number constant) | Percentage constant | Percentage falls (absolute number constant) |
| 2 First child whether son or daughter | Percentage falls rapidly towards zero (concentration on one family) | | Percentage falls slowly towards zero (concentration on one family) | |
| 3 All sons (*or* all daughters) | Percentage constant | | Percentage constant | |
| 4 All children whether sons or daughters | Percentage constant | | Percentage rises towards 100 per cent (equality of ownership) | |

In the first row of the table we consider the case in which parents always leave their property to the eldest son. In this case the absolute number of property owners each owning an unchanged amount of property will remain unchanged, since each property owner leaves it all to one son, who leaves it all to one son, and so on *ad infinitum*. In a constant population the percentage of families owning property will, therefore, also remain constant. But in a growing population the constant number of property owners will come to represent a smaller and smaller proportion of the population, as all sons after the first son in each family join the growing ranks of those without property. The analysis would be exactly the same if all families always left all their property to the eldest daughter instead of the eldest son.

In the second row we consider as an instructive intellectual exercise what is probably an unusual set of laws and customs, namely that the whole property is left exclusively to the eldest child whether a boy or a girl. In this case whether the population be stationary or growing the ultimate outcome will be for the whole property of the community to be owned by one single individual. Two properties can in this case be joined together in holy matrimony, but once joined they can never be separated, since death does not part them. If an eldest daughter with a property marries an eldest son with a property, this becomes a single property which will be left to the eldest child of the marriage. If that child marries a propertyless spouse, the enlarged property remains unchanged; but if he or she in turn marries a propertied spouse, then the already enlarged property is enlarged still further into a still bigger single property.

This process of concentration will continue indefinitely; but the speed with which it occurs will depend upon the degree of assortative mating. If there were perfect assortative mating among property holders, there would be a tendency for the number of property holders to be halved in each generation, since at every generation a male property and a female property would be merged into a single property. If mating were perfectly random, the process of property meeting property would be much slower. But the inexorable final result would be the complete concentration of all properties into a single ownership.

In row 3 of the table I consider the case where only men own property but where, unlike row 1, the property is divided equally among all sons instead of being left only to the eldest son. In the case of the stationary population where each father has only one son, the effect in row 3 is identical with the effect in row 1. But where the population is growing there is a difference between rows 1 and 3. Where only eldest sons inherit, the absolute number of families owning property will remain the same. Where all sons inherit, and where propertied and propertyless families are growing at the same rate, the percentage of families owning property will remain unchanged at its original 5 per cent. Once again the analysis would be unchanged if it was the daughters and not the sons who inherited the whole of the family property.

Neither in row 1 nor in row 3 does the degree of assortative mating have any effect upon the result. Indeed, the degree of assortative mating is in these cases meaningless; since either all women or all men are propertyless, there is no meaning to be attached to the degree to which men and women select spouses with properties similar to their own.

In row 4, however, the absence of perfect assortative mating is crucial. We consider now the case where properties are split up equally among all children, whether they be sons or daughters. If there were perfect assortative mating, properties would remain in the ownership of a privileged 5 per cent of the population, as in row 3. It makes no difference whether a property is left only to the sons in a family, or whether it is left half to the sons and half to the daughters, provided that these sons and daughters take as spouses the similarly endowed daughters and sons of similarly propertied families. Whether a whole property passes from a father to his sons who then marry propertyless wives or whether a half property passes to his sons who then marry wives who have received a similar share of a similar half property makes no difference to the property which they can then hand on to their children.

But if mating is not perfectly assortative, the difference between rows 3 and 4 is decisive. When properties are divided equally between sons and daughters and when the propertied sons may marry the daughters of propertyless parents and the propertied daughters may marry the sons of propertyless families, properties will be spread over a larger and larger number of the population. In the end there will result a complete equalization of property ownership. If any properties of unequal size remained, sooner or later they would meet, marry, and be averaged before being left to the next generation. Inequalities could thus be reduced; they could never be reintroduced. The smaller the degree of assortative mating, the quicker the process of equalization.

I need hardly add that laws and customs relating to inheritance do not consist exclusively of one or other of these pure forms. Moreover, of course, in the real world inequalities would be reintroduced and maintained by the accumulation of new properties and by all those factors of what I have called luck which lead to a dispersion about the average as new properties are accumulated; and the higher the degree of assortative mating according to properties, the greater the ultimate degree of inequality that will be sustained. In my artificial, mechanistic model of inheritance, I concentrated on a limited number of pure rules of inheritance, assumed that no new properties were accumulated, and omitted all the factors of dispersion which tend to restore inequalities solely to give an intuitive idea of the important underlying forces which over time the laws and customs of inheritance may be exerting in the background in society.

I introduced this discussion of the importance of laws and customs relating to the inheritance of property by pointing out that while parents could control the distribution of their property among their children, they could not control the distribution of their genes. There remains another very important reason for distinguishing between genetic inheritance and the inheritance of property. If Tom and Mary Jones decide to leave all their property to little Richard, they cannot leave any to little Jane as well. But if little Richard is lucky in the genes which he receives from his mother and father, this in no way reduces the chances of little Jane being equally lucky in the genetic draw. Or to put this in a somewhat different way, if a set of parents have four instead of two children, they can leave each child only one-quarter instead of one-half of their combined property; but they can endow each child with the same average genetic make-up, however few or however many children they may have.

This distinction is of fundamental importance when we consider the effects of differential fertility upon inequalities of income and wealth.

Suppose that the fertility of the fortunate were to rise and that of the unfortunate were to fall. As I have already pointed out, the fortunate parents would probably be able to accumulate somewhat smaller properties since they would have to support more children (line 49 in my diagram) and, on the assumption that the custom was to leave property equally divided among all children in the family, these somewhat smaller properties would be split into a larger number of fragments (line 19′ in Figure 9.1). Thus if parents have more children, each child can inherit a smaller share of what is probably a smaller total property. Conversely, the less fortunate families having a smaller number of children to support might be able to accumulate somewhat larger properties, and in any case, whatever properties they did accumulate would be less liable to be split into small fragments on the death of the parents. The effect of the differential fertility would undoubtedly be to mitigate inequalities in the ownership of property.

But there would be no such tendency to equalize genetic endowment. Having a large number of children in no way diminishes a parent's total genetic stock nor does it mean that this stock must be split into smaller fragments. An increase in the fertility of the fortunate relative to that of the unfortunate may raise the average quality of genetic endowments. But to equalize genetic endowment one would need to reduce the fertility both of the exceptionally fortunate and of the exceptionally unfortunate relative to the fertility of those with average fortune.

Endowments in social contacts probably fall in this respect somewhere in between genetic and property endowments. There are certain elements of social contact and atmosphere in the home which, like genetic endowments, can be enjoyed by all the children in the family, however few or many they may be. There are others, like expenditure on educational and similar social contacts, which, like property, if spent on Richard cannot be spent on Jane. There are still other elements which are intermediate; to have four instead of two children probably means that each child gets somewhat less attention, but more than half as much, from his parents.

To conclude, my remarks on the various relationships which determine the transmission of personal endowments have, I fear, been rather disjointed; but I hope that I have said enough to make it clear that they are all interrelated in a rather complicated single biological–demographic–social–economic system.

In any case the analysis remains woefully incomplete unless one can estimate quantitatively the relative importance of the various factors. The difficulties of quantifying the relationships are immense. First, it is extremely difficult to get measurements of many of the relevant variables. For example, genetic endowment may affect many hitherto unmeasured characteristics which are economically much more important than the IQ scores which we can measure and which for that very reason have been so much examined. Second, the very marked correlation between the various components of good and bad fortune which I have emphasized in this lecture itself makes quantitative measurement of the separate importance of each component statistically very difficult. Third, the very complexity of the intertwining of so many genetic, demographic, sociological, and economic factors raises very formidable problems for empirical research in this field.

In their recent book entitled *Inequality* Professor Jencks and his colleagues at Harvard claim to have shown that the factors which I have called luck are immensely more important in the explanation of inequalities between individuals than the structured biological, demographic, social, and economic factors which I have called fortune and on which I have concentrated in this paper (Jencks 1972).

This may well be so; and if it is so, it has very far-reaching implications for the design of policies if we want to reduce inequalities. Many people and not only Marxists have maintained that we must rely more on structural changes in society's institutions which will basically readjust what I have called the structural endowments of good or bad fortune. But if Professor Jencks is correct, we should on the contrary rely less on factors of educational, social and economic reform which will equalize people's structured fortunes in life and more on a continuing direct day-to-day redistribution of the unequal incomes and properties which the chances of luck will continually be re-establishing in society. Such measures – for example, progressive taxation of incomes and property, negative income taxes, social dividends and other social benefits, minimum wage rates, free education and medicine – would be needed simply because of their immediate direct effect on the standards of the lucky and the unlucky within any one generation.

Perhaps in the present state of our knowledge we should put more emphasis on such direct measures. If Professor Jencks is correct, that is the only way. If he is incorrect, such measures, in addition to their immediate impact effect on the equalization of incomes and property within any one generation, will also help to set in motion in the right direction many of the self-reinforcing influences in society which I have catalogued in this lecture, since more equal incomes and properties may lead to somewhat more equal educational, social and economic opportunities and thus, for what it is worth, to a more equal intergenerational transmission of endowments.

But I confess that I am disinclined to rush to this conclusion. I understand that the results of the many valuable empirical studies on these matters which have been and are being conducted are still to a considerable degree uncertain, controversial, and sometimes inconsistent with each other. There remains the possibility that fortune is not quite so secondary to luck as Professor Jencks considers it to be.

Thus something may still be gained from considering carefully those factors whose importance Professor Jencks is denying; for they really do at first sight appear to be very influential factors. Indeed, I must confess that I do find Professor Jencks's conclusions very surprising, although I have not the competence to criticize his sophisticated, careful and scholarly statistical work.

I have already chosen an epitaph for inscription on my tombstone, namely: 'He tried in his time to be an economist; but common sense would keep breaking in.' It certainly is a useful irruption when common sense breaks into a sophisticated economic model to point out that the assumptions of the model simply rule out all the factors which casual empiricism suggests are important in the real world. But I cannot apply my common sense in that manner to Professor Jencks's work which covers comprehensively, but finds unimportant, all the main factors which my casual empiricism suggests to be important. Alas, common sense may imply no more than the conservative conventional wisdom which refuses to face new hard facts because they are disturbing. I know that in the end I must face the facts; but meanwhile I am surrendering to my common sense to the extent of preserving an open mind for just a little longer.

# 10 Social class structure and the genetic basis of intelligence

## B. K. Eckland

(From R. Cancro (ed.), *Intelligence: Genetic and Environmental Influences*. New York, Grune & Stratton, 1971. pp. 65–76. Reprinted by permission of Grune & Stratton Inc. and the author.)

The present chapter deals with a proposition which some of us take for granted, to which some would take strong exception, and perhaps to which most of us just have not given much thought. The proposition is that in meritocratic societies the average difference in measured intelligence of children from different social classes has partly a hereditary basis. In other words, talent (genotypically) is not distributed randomly in each new generation across the entire class structure but tends to be disproportionately concentrated at the middle and upper levels. Middle class children, *on the average*, tend to be innately brighter than lower class children.

Before examining the proposition and its implications, let me make at least two concessions to those who are inclined to challenge these statements or would prefer to ignore them: (1) intelligence is not fixed at birth or at conception, since many different factors, often in combination, contribute to its development; and (2) owing to imperfect assortative mating for intelligence and sociocultural differentials, many lower class children are much brighter than some middle class children. Perhaps too much has been made of the point that genes do not fix the development of polygenic traits like intelligence. Many different kinds of environmental factors are involved. Nevertheless, genes are discrete entities, which in combination determine the limits of the range within which their phenotype will be expressed. The idea of an unlimited range of individual intelligence modified only by environmental factors just does not fit with reality.

Because both nature and nurture are involved and because they are never perfectly correlated, there is a considerable amount of overlap in any group distribution of intelligence. No one class has a monopoly on all the 'good' genes and another on all the 'bad' genes. Both low to moderate assortative mating for intelligence and mild to severe environmental deprivation guarantee a relatively substantial pool of very bright children from lower class backgrounds. But, as we shall argue later, the amount of overlap between classes tends to diminish as assortative mating and equality of opportunity both increase.

### The deprivation model

In the standard deprivation model of social class and intelligence, any observed correlation between the social status of an adult population and the intelligence of its children is explained in terms of intervening environmental factors that enhance or inhibit cognitive development and performance, factors which themselves are correlated with social status.

The model is depicted in Figure 10.1 by a broken arrow in the path between intelligence and social status, $r_{IS}$, and causal arrows in the paths connecting intelligence with environment, $p_{IE}$, and environment with social status, $p_{ES}$. We also have introduced a residual factor, $R_e$, which symbolizes all possible determinants of the child's environment unaccounted for by parental status.

The extent to which children's intelligence is correlated with the socioeconomic status of their parents, of course, depends upon time and place as well as upon the age of the children. Nevertheless, most of the correlations reported in the literature appear to fall in the region between 0·35 and 0·40 and, according to Jensen (1969), this 'constitutes one of the most substantial and least disputed facts in psychology and education' (p. 75).

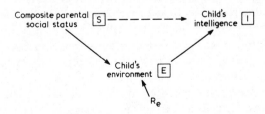

*Fig. 10.1* Standard deprivation model of social class and intelligence

There are three basic types of environmental variables applicable to the deprivation model that may help account for these findings. One group of variables is *physical* or *biological* in nature, such as prenatal care, nutrition and birth order. Severe prenatal stress and malnutrition both appear to impair cognitive development, and each of these conditions is found more frequently among the disadvantaged segments of the population. Birth order also may be partly responsible for the class-intelligence correlation. For whatever reasons, first-born children on the average score slightly higher on mental tests; and, owing to below average rates of fertility, there are proportionately somewhat more first-born children among higher socioeconomic groups.

A second set of correlates of social class that sometimes contribute to the variance in mental ability is *cultural*. Social classes can be viewed as 'subcultures' in the sense that they confer upon their members an integrated set of norms and values which are carried down from generation to generation. The segregation, degradation, and continuity of the poor serve to insulate them from the mainstream of society. As a separate culture, the poor generally experience a relatively unique pattern of behavioural and psychological traits which may directly impair the social, emotional, and cognitive development of children raised under these conditions (Eckland and Kent 1968).

The third general group of environmental variables that may help account for some portion of the class-intelligence correlation is *social structural,* usually defined as differential access to the institutionalized means for achieving culturally prescribed goals (Merton 1957). Apart from the content of the subculture of poverty, deprivation frequently involves socially structured inequalities in education and other opportunities for improving a child's performance. These inequalities usually are attributed to the control one class exerts over another, i.e. the superordinate element benefits by keeping another element dependent.

Thus, numerous class-linked environmental variables have been found to contribute to individual differences in measured intelligence. Unfortunately, no serious attempt has ever been made to estimate in a direct fashion the additive effects of more than a few of these factors at one time. In what little that has been done, the effects have not been found to be strictly additive. The 'independent' variables themselves are highly interrelated. But even if this were not the case, the cumulative effects of environmental variables probably would not wholly explain social class differences in intelligence since there is another set of factors which the deprivation model totally ignores.

## The polygenic model

If native talent is not randomly distributed among all children at birth in the class structure, then the correlation $r_{IS}$ cannot be entirely the result of cultural and other forms of environmental deprivation. The polygenic model in Figure 10.2 holds that *both* environmental and hereditarian variables are required to explain social class differences in intelligence. As long as a bona fide argument can be made that $p_{SM}$, $p_{MP}$, $p_{HP}$, and $p_{IH}$ are all greater than zero, then the product of these paths constitutes a genetic loop which partly accounts for the correlation $r_{IS}$.

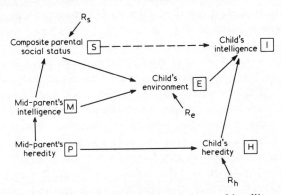

*Fig. 10.2.* Simplified polygenic model of social class and intelligence.

Three new variables have been introduced in Figure 10.2. Mid-parent's and child's *heredities* refer to the genotypes of intelligence or to that particular set of polygenes that produce quantitative variations in cognitive functioning. We have used *mid*-parent's heredity and *mid*-parent's intelligence only as a matter of simplification. Mid-parent's and child's *intelligence, I* and *M*, both, of course, refer to the phenotypes of intelligence or, in other words, whatever IQ tests generally measure.

We also have introduced two new residuals in the model. $R_s$ stands for all sources of variance in the parents' composite social status unaccounted for by their intelligence. $R_h$ stands for all sources of variance in the child's heredity unaccounted for by the mid-parent's heredity.

Let us now examine each of the main paths or links in the genetic loop. Although along the way we shall note some figures suggested by the literature, no claim will be made here regarding the true magnitude of any of the links. We need only show that a connection actually exists between each variable in the loop in order to support the model. Nevertheless, the weight of our argument obviously does depend upon their strength, and one very weak link would seriously limit our conclusions. This is because any coefficients entered in the paths are multiplicative and not additive, which also means that a small change in one of the weaker paths would contribute more to the overall effects of the genetic loop than a comparable change in one of its stronger paths.

### Parental intelligence and social status

The path from parental intelligence to social status, $p_{SM}$, is the most critical because it is the primary link between the genetic and environmental components of the model and also because it appears to be the weakest link in the genetic loop. The zero-order correlations found in most studies range from 0·20 to 0·50, i.e. when social status is measured in terms of occupational prestige.

Although few will deny that intelligence is a bona fide cause of adult achievement, it is not entirely clear just how this operates owing to the confounding effects of education. When years of schooling are taken into account, only a small empirical relationship remains between intelligence and occupational achievement (Eckland 1965, Bajema 1968). This should not be interpreted, however, to mean that the relationship is spurious. Rather, it probably simply means that the educational system acts as the central mechanism in the identification of talent and the allocation of status. Success in school or college depends upon mental ability, while one's point of entry in the labour force and subsequent job promotions depend, in turn, upon prior success in school or college.

Let us inspect each of these steps, especially the first, more closely. The relationship between intelligence and educational achievement, of course, should not be surprising since IQ tests originally were designed to predict academic performance. Student performance in the classroom presumably depends, among other things, upon the kinds of mental skills which IQ tests measure. There is much evidence that they usually do (Goslin 1963, Lavin 1965). Yet, aside from the true validity of these tests, the association between intelligence and educational attainment is built into the educational selection process itself. That is, standardized tests of mental ability are employed as a primary basis for sorting and selecting students beginning in the primary grades and continuing on through college. In 1967 I summarized this development as follows:

(a) Virtually all schools now use standardized tests. Moreover, a large majority of the secondary schools in this country plan to expand their testing programmes in the near future.

(b) Standardized tests are used to assess the potential learning ability of students, in order to provide individualized instruction, and to guide students in their decisions about school curricula, going to college, and jobs. A fairly large majority of the nation's youth actually are being placed into homogeneous classes on the basis of these tests, either by establishing completely separate programmes of instruction for different students ('tracking') or by assigning students to different sections of the same course ('grouping').

(c) A majority of both students and adults believe that standardized tests measure the intelligence a person is born with; although most, at the same time, recognize that learning makes an important difference. Students and parents alike believe that the tests are basically accurate. While teachers essentially agree with the students and parents on these points, they also tend to believe that the tests are their best single index of a student's intellectual ability.

(d) Although public acceptance of standardized testing seems to depend largely upon the specific purposes for which the tests are used, a majority of the students believe that these tests are and should be important, especially in terms of deciding who should go to college.

This is only part of the story. For the college-going aspirants, the first two major hurdles – Educational Testing Service's Preliminary Scholastic Aptitude Test (PSAT) and National Merit Scholarship Corporation's Qualifying Test – usually come in the eleventh grade. Scores on the PSAT are not used to decide who goes to college, since that decision presumably was made some years earlier. Rather, one of the primary purposes of the test is to help decide who goes where to college by giving each student an early estimate of the probabilities of being

admitted to any particular school. The National Merit examination, on the other hand, is designed to select among the
uppermost 2 per cent of the country's students those who will
receive special commendation and awards. While in competition
with other programmes, the growth of the PSAT and National
Merit programs in the past ten years has been so phenomenal
that both are being administered today in roughly three-fourths
of the nation's high schools.

The next major, and perhaps most important, challenge
comes about a year later – the college admissions tests and, for
those in accelerated programmes, the Advanced Placement exam-
inations. Whether administered nationally or locally, tests are
being used today by most colleges in their admissions process or
for placement. Indeed, the proportion of high-scoring students
that a college can attract has become the most objective criterion
available for ranking the colleges themselves. In addition, many
secondary schools and colleges are participating in the Advanced
Placement Programme of the College Entrance Examination
Board. Although presently involving fewer students and
schools, its growth has closely paralleled that of the National
Merit programme.

As the student progresses further, the significance of educational testing does not diminish. Upon completion of most
undergraduate programmes, admission to nearly any reputable
graduate or professional school is fast becoming dependent, in
part, upon the student's sophistication on such nationally
administered tests as the Graduate Record Examination, the
Admission Test for Graduate Study in Business, or the Law
School Admissions Test. Likewise, entry into a particular field
direct from a four-year undergraduate programme sometimes
requires additional testing, such as the National Teachers
Examination or the Foreign Service Examination. Even the
Peace Corps requires the Peace Corps Entrance Tests. (Eckland
1967, pp. 185–7)

Thus, the use of IQ and closely related tests of ability can now
be found at nearly every point in the educational system. In fact,
as the testing movement has gained momentum, many observers
are wondering if our schools place *too much* emphasis on these
programmes, particularly when it is recognized that culturally
biased tests tend to discriminate against underprivileged groups.

Yet, notwithstanding even the restrictive legislation which has been enacted recently in some communities in order to safeguard students against the arbitrary use of tests, school and college testing programmes no doubt will survive and perhaps play a still greater role in the future. I doubt that we ultimately shall decide who goes where to college or similar educational decisions by lottery, as some have suggested that we ought to do. Lotteries probably are no more 'equitable' than testing programmes, and they certainly are a less rational way to use human resources – although we recognize too that equal educational opportunity in its purest form might be a more compelling objective in the long run than the efficient utilization of talent.

Probably stronger than the relationship between intelligence and educational achievement is the relationship between educational and occupational achievement (occupational status is the most commonly used index of social status). Commercial and industrial enterprises seem to rely very heavily upon our schools and colleges not only to train people, but also to sort out those who are less motivated and inept. Although other factors are involved too, getting to the head of the academic procession takes both persistence and 'brains', and it is quite clear that graduates of the nation's best colleges and universities tend to get the most and the best job offers. Thus, just as the holder of a degree from any four-year college is almost guaranteed passage into middle class status, graduation from a highly selective and prestigious college or university is a reasonably valid passport for entry into upper middle class status.

This all means that as members of the younger generation graduate from school or college, enter the labour force, and have families of their own, the relationship between intelligence and adult social status (in our model) should at least retain its present strength and quite probably become a bit stronger. The expansion of most of the school and college testing programmes described earlier has been too recent to have had much effect on the vast majority of American adults. Only the most recent cohorts of students are being affected. As they move into the labour force, we should expect the path $p_{SM}$ to increase.

*The inheritance of intelligence*
The next link in the genetic loop, $p_{MP}$, and the last link, $p_{IH}$, will be discussed together. In both paths we are dealing with the

question: 'What proportion of the variance in measured intelligence is due to heredity?' This proportion usually is expressed in terms of a heritability coefficient, and its magnitude will be approximately the same whether we are speaking of the child's or the parents' intelligence.

Much controversy has revolved around the concept of heritability and its measurement. But the most serious issues probably can be avoided if we remember that in the strictest sense a heritability measure is a property of a population, not of an individual or of the trait itself. Heritability simply estimates the proportion of the total phenotypic variance in a particular population at a particular point in time. We have no direct measure of heritability for individuals, *per se*. Moreover, a high heritability coefficient does not mean that the environment is unimportant or unnecessary for the expression of the trait. Rather, it would only mean that environmental variation does not explain very much of the *variation* in the trait that is being observed.

Heritabilities for intelligence generally fall within the range from 0·70 to 0·90, their size depending upon the population under consideration, plus the particular methods, formula and tests used.

It could be argued that the heritability of intelligence is increasing. Since the coefficient is a population statistic and always depends upon the absolute amount of variance of trait-relevant factors in the environment, any significant change in the environmental factors will change paths $p_{MP}$ and $p_{IH}$. The size of these paths is inversely related to the total amount of environmental variance. This means that if either of the two direct paths from parental social status ($S$) and parents' intelligence ($M$) to the child's environment ($E$) declines, then, all other things being equal, we should expect an increase in the proportion of the variance in measured intelligence that can be attributed to heredity. To the extent that this nation's efforts to upgrade the environments of low *SES* children are successful, this should be exactly what happens. Since the genetic and environmental factors influencing intelligence presently covary in the same direction, successful efforts to decrease the excess number of children generally found in the lower region of most IQ distributions also would tend to decrease the absolute amount of variance observed and shift the population mean upward.

However, there are and would continue to be other environmental factors (in the residual, $R_e$) affecting cognitive development that are uncorrelated with either parental status or parental intelligence. If parents gave up altogether the care of their children to the schools or some other complex of socializing agencies, one of the major sources of environmental variation would have disappeared. On the other hand, institutionalized child-rearing could conceivably create more rather than less overall diversity of trait-relevant environments, especially if highly individualized treatments of the kind being suggested by some educators were put into practice. If this were the case, heritabilities might decrease instead of increase, and the total population variance might increase instead of decrease.

There is a further complication that probably would occur. If different treatments were assigned routinely to different individuals, it is unlikely that they would be assigned randomly. The 'richness' of each child's experience instead may vary directly with the unique package of genes (or potential) he brings to the situation. Just what this would do to the size and meaning of heritability has not been adequately explored.

Nevertheless, as far as our model is concerned, current heritabilities of intelligence are quite high and contribute substantially to the genetic loop.

### The genetic parent-child correlation

The final link to be considered in the loop, $p_{HP}$, also is quite strong. And, like the other links, its size depends upon certain environmental conditions.

A child receives half of his genes from each parent. Thus, at the minimum, the genetic parent–child correlation for intelligence is 0·50. However, it apparently is considerably larger since this figure only applies under conditions of random mating. The additive genetic parent–child correlation for intelligence and other polygenic traits is actually 0·50 plus one-half the assortative mating coefficient or, in other words, the degree to which parents on the average hold certain genes in common. The more closely the parents resemble each other, the more closely children will resemble their parents. Assortative mating also tends to increase a trait's average homozygosity and, thus, the population variance of the genes, which in the long run means reducing the trait's within-class variance while increasing its between-class variance.

Unfortunately, the size of the assortative mating coefficient for intelligence – and, consequently, the genetic parent–child correlation – is somewhat a matter of conjecture, especially for recent cohorts. One problem is that only phenotypic, not genotypic, correlations are available. Nevertheless, if the coefficient of assortative mating is as high as 0·60, as some authors have suggested (Jensen 1969), then the parent–child correlation, $p_{HP}$, must be on the order of 0·80.

One possible check on the size of this path is to use some of the values that have been suggested and see if the product of the three paths, $p_{IH}p_{HP}p_{MP}$, equals the *phenotypic* parent–child correlations commonly reported in the literature. Thus, if heritability is 0·80 in both generations and the genetic parent–child correlation also is 0·80, the product of these paths is 0·51, which is remarkably similar to the parent–child correlation ($r_{IM}$) usually observed (Erlenmeyer-Kimling and Jarvik 1963). Under conditions of random mating the expected value for this correlation would be only 0·32.

Due largely to educational selection and to changes in social values regarding mate selection, assortative mating for intelligence probably is increasing. Societies that value individual freedom of choice in marriage and at the same time place potential mates into relatively homogeneous social settings based on ability are likely to produce high rates of assortative mating for intelligence. Males and females of like intelligence, especially in recent generations, generally end up in similar educational niches, such as school drop-outs, as college classmates, or as graduate students. These set the broad limits within which mate selection tends to occur.

Let me give one example. In 1965 a follow-up study was conducted by the National Merit Scholarship Corporation to determine the accomplishments of all female Merit Scholars who had received their awards between 1956 and 1960. These women undoubtedly were among the brightest in the nation. They were not destined to be spinsters. By 1965 the majority had married and most of their husbands had either completed or entered graduate school (Watley 1969). Since their husbands no doubt also are far brighter than average, their children will be well above average, and this partly will be due to the hereditary basis of intelligence.

If not already the case, education may soon become a better predictor of who mates with whom than either social class or residential propinquity. We already have argued that educational testing and similar mechanisms create relatively homogeneous pools of talent. We only need to add that our schools and colleges also function as marriage markets. I suspect that if they did not, the college-going rates for women in this country would be markedly lower than they are. But, as it is, the number of women attending college for the first time has been rising more rapidly over the past generation than the number of men attending for the first time. Yet, it is not entirely clear exactly what value a college diploma is to most women, other than attracting an eligible husband and unrestricted passage to middle class respectability (there are, of course, exceptions).

Once again, assortative mating increases the genetic parent–child correlation, which, in turn, strengthens the genetic loop. Since the four paths in the loop are multiplicative, their total effect presently accounts for perhaps only a relatively small proportion of the association between parental social status and a child's intelligence – even though individually some of the paths have quite strong coefficients. Nevertheless, the polygenic model of social class and intelligence, I believe, does operate in approximately the manner described here, and the strength of the genetic loop is sufficient to warrant closer inspection. This will become increasingly true if contemporary societies continue to implement the principles of the meritocracy – a trend that, in the long run, seems inescapable.

### The future of social mobility

In conclusion, I would like to consider the future of social mobility by projecting the trends outlined. Sociologists generally define social mobility as the intergenerational movement of individuals from the social class of their family of origin to the class of their family of procreation. Sometimes the individual's intelligence is considered, but only as one of several plausible mechanisms through which parental status limits or enhances the individual's future status. No genetic components of intelligence are recognized, and, like the deprivation model described earlier, the correlation between the status of father and son is usually thought to be wholly a result of social inequality. In fact, the size of the correlation sometimes is taken as a direct measure of inequality of opportunity – or the extent to which positions are socially ascribed rather than achieved in accordance with one's own effort and abilities.

However, the polygenic model of social class and intelligence suggests a quite different conclusion regarding the effects of equality of opportunity on social mobility. We already have argued that equal opportunity in the long run tends to strengthen most if not all the paths in the genetic loop between parental social status and child's intelligence. Thus, rather than leading to progressively lower correlations between father's and son's status, equal opportunity may eventually raise the correlation. Or, in other words, social mobility would not increase over time; rather, it would eventually either stabilize or decline.

In a completely open society under full equality of opportunity, a child's future position might be just as accurately predicted from the status of his biological parent as in a caste society. The basic difference between the two being that in a completely open system the causal links between generations would involve character specific or polygenic traits like intelligence, whereas in a caste system the links are consanguineous or cultural.

However, in the transition from castelike to truly open-class systems, it is not likely that a linear transformation in the rates of mobility would be observed. Initially, the correlations between father's and son's status should fall as meritocratic principles begin to operate (this is what we still seem to be observing today). But, at some optimal point in the not too distant future the force of the genetic loop may swing the correlations in the opposite direction.

When industrial societies emerged along with their complex division of labour, it eventually became reasonably clear that not everyone could do everything. Perhaps a noble's son could once be trusted to oversee a kingdom, but who is willing to trust the laws of primogeniture, nepotism or even seniority in the running of today's social machinery? Thus, with the development of mass education and the demand for skilled technicians and a managerial élite, who does what now depends more upon what people are capable of doing than who their parents happen to be or other irrelevant criteria.

But what of the future? If, as suggested, the heritability of intelligence remains high while educational selection and assortative mating for intelligence increase, what then? Have we already reached the turning point, in which case we should soon expect declining rates of mobility? Or, perhaps for a variety of reasons we shall never reach this point. These are, I believe, significant questions. Unfortunately, our data are far too weak to provide even the most tentative answers.

# 11 Genetics, social structure and intelligence

## A. H. Halsey [1]

(From *British Journal of Sociology*, 1958.)

## Introduction

It is our purpose in this paper to consider the question of social class differences in measured intelligence as an illustration of the more general problem of the relation between social structure and the distribution of genetic characteristics in populations. Though innate intelligence may legitimately be inferred from available evidence, the evidence is also strong that it is determined not by one gene but polygenically. On these assumptions a genetic model may be constructed which, when relevant variations in social structure are taken into account, throws considerable doubt on the possibility that mean differences between classes in measured intelligence are innate. This doubt arises from the fact that the relevant aspects of social structure have not been stable enough for long enough in differentiated societies to allow sufficient time for the kind of polygenic process involved in human intelligence to develop class differences of the order suggested by the known social distribution of measured intelligence. This conclusion is, moreover, strengthened by a consideration of the variance found in measured intelligence among social class groups.

## 1 The theory of intelligence

On the evidence, especially of correlation coefficients derived from studies of twins, siblings and foster children, it may be concluded that there is a genetic component in the intelligence test performance of individuals, and hence it is reasonable to postulate a genotype of innate intelligence which we shall call *IA* (Vernon 1955).[2]

The actual manifestation of intelligence, which we shall call *IB*, is a social product of adaptive behaviours within the framework laid down by *IA*. Largely on the basis of the work of Hebb and Piaget, Vernon describes *IB* as the set of behavioural character- istics correlative to Hebb's 'phase sequences' or patterns of discharge in the association areas of the brain. Phase sequences, and their counterparts in adaptive behaviour, are dependent for their formation on experience of appropriate environmental stimuli. Piaget too views intelligence as a developing complex of adaptive behaviours beginning with simple reflexes in the new-born child and ending with intricate conceptualizations and trains of reasoning in the adult. The growth of intelligence is dependent upon stimulation in social interaction, the use of language and the manipulation of objects. To say that *IB* is a complex is also to emphasize that intelligence is no one single factor. We do not know what are the factors involved. If we did then the problem of the relation between *IA* and *IB* could be more successfully tackled.

Our present knowledge of factors in intelligence is derived from intelligence tests. Performance in intelligence tests we shall call *IC*. Any one of the many measures of intelligence (*IC*) consists of a *sample* from the set of behaviours which constitute *IB*. It is a biased sample because of the use of intelligence tests for educational selection and their validation predominately in terms of educational success. The factors derived from analysis of intelligence tests are determined by the content of the tests and this in turn by the requirements of prediction of educational performance.

That the determination of *IB* and *IC* is partly social is well known, though our knowledge of the relevant social processes is unsystematic and incomplete. What, however, is much less fully appreciated is the important role played by social structure in the distribution of genetically determined characters among which the postulated *IA* is here considered as an example. A brief consi- deration of the mechanics of inheritance should make this clear.

*Genetic variation*

It is of the greatest importance to bear in mind that the appropriate unit for consideration of genetic variation is not the individual but the *population* of individuals between whom mating is possible. The stock of genes present in an endogamous population is usually called a gene pool. Limits to the variation of a population are set at any one time by the composition of its gene pool and these limits vary from one generation to another according to four factors: (1) genetic drift [3], (2) mutation, (3) selection, and (4) migration. For the purposes of the present problem the factors of importance are selection and migration.

Within populations, genetic variation between individuals is ensured by heterozygotes, i.e. cells carrying alternative (allelic) genes. In fact, only a small proportion of the possible gene combinations in any species is ever realized. With $n$ heterozygous genes there may be produced $2^n$ kinds of sex cells with different gene combinations. For example, an individual with twenty heterozygous genes may produce 1,048,576 kinds of sex cells. Thus, since most people are heterozygous for thirty or more genes, it becomes 'a reasonable guess that no two persons alive (identical twins excepted) carry the same genes' (Dobzhansky 1955, p.34).

Heterozygotes increase the potential variability of a group over that actually exhibited by its members and this adaptive advantage is greater the faster the breeding rate, i.e. the more rapid the rate of sampling of the possible gene combinations in the gene pool. Phenotypic variation is, in general, an outcome of cumulative interplay between these factors and the accidents of environment.

*Social structure and genetic variability*
There are sociological aspects of genetic drift, mutation, selection and migration. One important illustration may be taken from the pattern of human social life for the greater part of the history of man. This pattern, having been one of small scattered endogamous groups, has tended to maximize genetic drift, because of isolation and small numbers, and to form trial gene pools for mutations. However, in the modern world these conditions for rapid evolution or differentiation have passed away with the formation of vast states and nations and hence large endogamous populations. In essence this is the basis of Sir Arthur Keith's group theory of human evolution.

Again, to take a second illustration, some genotypes are more viable, more socially attractive and more fecund than others. The resultant differential fertility is partly determined by social structure. Sexual selection and control of migration (social and geographical mobility) are found in all human mating systems – never panmixia. There are universal rules of exogamy and widespread rules and practices of endogamy. The former lead to increased frequency of heterozygote individuals and the latter to increased frequency of homozygotes. Thus, though deviations from panmixia do not usually result in alterations in the gene frequencies of the total population, the facts of social differentiation produce a tendency towards homozygous genetic clusterings within societies.

These two examples are sufficient to indicate that the course of genetic evolution and of variations within the human species are limited and partially controlled by variations in social structure, especially the size of populations, the degree of endogamy and exogamy in their mating systems and the ease of geographical migration and social mobility between groups within them. The postulated character of innate intelligence ($IA$) may now be considered from this point of view.

## 2 A genetic model of intelligence

We begin with an assumption of single gene determination of $IA$. This is a gross oversimplification but nonetheless instructive because, in any given social structure, it gives the maximum weight to genetic factors in the selection process which we are about to discuss.

Let it be assumed that $IA$ is produced by inheritance through a gene $A$ such that the homozygote $AA$ has the value of 120, the heterozygote $Aa$ the value of 100, and the homozygote $aa$ the value of 80.

### Case 1: a caste system

We may begin consideration of the effect of variations in social structure on the social distribution of innate intelligence ($IA$) by imagining a society having a population of 1000 divided into two castes – a low caste numbering 900 and a high caste numbering 100. Castes are by definition endogamous and let us further assume that there is random mating within them and equal fertility between them.

Looked at genetically and ignoring mutation and genetic drift [4], this society consists of two separate gene pools, each in Hardy-Weinberg equilibrium, the Hardy-Weinberg formula $[p^2 + 2p(1-p) + (1-p)^2]$ being a mathematical statement of the frequency ($p$) of a genotype in a random mating population in successive generations. Thus, if the frequency of the gene for high intelligence $(A)$ in the low caste is $0.3$ $(pA^L = 0.3)$ and in the high caste is $0.7$ $(pA^H = 0.7)$, the following equilibrium results:

|  | Low caste | | | | High caste | |
|---|---|---|---|---|---|---|
| Proportions: | $0.49aa$ | $0.42Aa$ | $0.09AA$ | $0.09aa$ | $0.42Aa$ | $0.49AA$ |
| Numbers | $441aa$ | $378Aa$ | $81AA$ | $9aa$ | $42Aa$ | $49AA$ |
| Means: | (Low caste) $I_A{}^L = 92$ | | (Total population) $I_A{}^T = 94$ | | (High caste) $I_A{}^H = 108$ | |

In words, the result under these genetic and social conditions is that:

1 The high caste has the higher mean (108 compared with 92).
2 The high caste has a much higher proportion but a much lower number of individuals with high intelligence (49 compared with 81).
3 There is a considerable overlap between the castes in the distribution of $IA$, and, genetically, this means that there are many more $A$ genes in the low than in the high caste, most of them in heterozygous conditions.

The situation described would remain stable between generations. It should, however, be noticed that the means of $IA$ are a function of the frequencies $(pA)$ chosen in the example. We could, for instance, have shown equal $pA$'s in the two castes and in that case the mean values of $I_A{}^L$, $I_A{}^T$ and $I_A{}^H$ would have been equal.

*Case 2: a class system*
The assumptions made in the caste model could not be found in any actual society and are certainly very far from representing the kind of social structure to be found in modern Britain. We may move closer to the latter by imagining a society in which social position is determined by selection according to $IA$. In the example which follows we continue to ignore mutation and genetic drift, we retain our assumption that innate intelligence is determined by a single gene but we introduce new migration (or mobility) assumptions. We postulate a low class of 900 and a high class of 100 with, in each generation before marriage, mobility upward of the ten most intelligent members of the low class and mobility downward of the ten least intelligent members of the

high class. Mating is again assumed to be random within classes and fertility equal between classes.

The process begins at generation $F_0$ with gene frequencies $pA$ and $pa$ equal (i.e. 0·5) in both classes. Then the proportions of $aa$, $Aa$ and $AA$ individuals are 0·25, 0·50 and 0·25 respectively in both classes and the corresponding numbers of individuals are $225aa$, $450Aa$ and $225AA$ in the low class and $25aa$, $50Aa$ and $25AA$ in the high class. Mobility then takes place to alter the distribution of individuals so that there are $235aa$, $450Aa$ and $215AA$ in the low class and $15aa$, $50Aa$ and $35AA$ in the high class. The gene frequencies become, at the same time $pa^L = 0·5111$ and $pa^{H} = 0·4000$ so that, after mating, the distribution of individuals in generation $F_1$ is, for the low class, $235aa$, $450Aa$ and $215AA$ and, for the high class, $16aa$, $48Aa$ and $36AA$. This distribution gives the low class a mean intelligence $(I_A^L)$ of 99·56 and the high class $(I_A^H)$ 104. A summary of the process, continued for subsequent generations is given in Table 11.1, and frequencies of the gene for low intelligence $(pa)$ and of the mean innate intelligence scores $(IA)$.

It may be seen from Table 11.1 that the trend of the distribution of $IA$ has the following characteristics:

1  The mean $I_A^L$ decreases at a decreasing rate to 98; the mean $I_A^H$ increases at a decreasing rate to 120 and the mean $I_H^T$ is constant at 100.
2  The class distributions continue to overlap, the majority of people with high intelligence still remaining in the low class.
3  Only over seven generations would low intelligence $(a)$ be bred out of the high class. High intelligence would never be bred out of the low class. At the eighth generation the social distribution of $IA$ stabilizes with a constant rate of 'circulation' of $AA$ between the classes.
4  The number of homozygotes in the total population increases in each generation (from $F_1$ to $F_7$), i.e. there are more people of high intelligence *and* more people of low intelligence, i.e. the standard deviations increase in each generation.
5  In other words, the heterozygotes form a reservoir of both high and low ability which, in this case, is drained high to high class and low to low class up to the eighth generation.

Thus the general effect of selective migration through intelligence, other things being equal, is to increase the frequency of

Table 11.1 Social distribution of intelligence (IA) in a class society with single gene inheritance of intelligence

| Genera-tion | Low class | | | | | | High class | | | | | | Mean intel-ligence of total population |
| | Gene frequency | | Number of persons | | | Mean intel-ligence | Gene frequency | | Number of persons | | | Mean intel-ligence | |
| | $pa^L$ | $pA^L$ | $aa$ | $Aa$ | $AA$ | $I_A^L$ | $pa^N$ | $p_A^H$ | $aa$ | $Aa$ | $AA$ | $I_A^H$ | $I_A^T$ |
| | (1) | (2) | (3) | (4) | (5) | (6) | (7) | (8) | (9) | (10) | (11) | (12) | (13) |
| $F_0$ | 0·500 | 0·500 | 225 | 450 | 225 | 100·0 | 0·500 | 0·500 | 25 | 50 | 25 | 100·0 | 100 |
| $F_1$ | 0·511 | 0·489 | 235 | 450 | 215 | 99·6 | 0·400 | 0·600 | 16 | 48 | 36 | 104·0 | 100 |
| $F_2$ | 0·522 | 0·478 | 245 | 449 | 206 | 99·1 | 0·300 | 0·700 | 9 | 42 | 49 | 108·0 | 100 |
| $F_3$ | 0·533 | 0·467 | 256 | 448 | 197 | 98·7 | 0·205 | 0·795 | 4 | 33 | 63 | 111·8 | 100 |
| $F_4$ | 0·541 | 0·459 | 263 | 447 | 190 | 98·4 | 0·134 | 0·866 | 2 | 23 | 75 | 114·6 | 100 |
| $F_5$ | 0·547 | 0·453 | 270 | 446 | 186 | 98·1 | 0·075 | 0·925 | 1 | 14 | 86 | 117·0 | 100 |
| $F_6$ | 0·553 | 0·447 | 275 | 445 | 180 | 97·9 | 0·022 | 0·978 | 1 | 4 | 96 | 119·1 | 100 |
| $F_7$ | 0·556 | 0·444 | 278 | 444 | 178 | 97·8 | 0·000 | 1·000 | 0 | 0 | 100 | 120·0 | 100 |

Note: For purposes of presentation numbers are rounded to whole numbers in columns 3, 4, 5, 9, 10 and 11, to one decimal place in columns 6 and 12, and to three decimal places in columns 1, 2, 7 and 8.

homozygotes and to decrease heterozygotes, i.e. to increase the incidence of both high and low intelligence and to reduce the incidence of middling intelligence.

These results have been obtained from a very simple model of society. If we wish to move still nearer to reality, many more complex assumptions as to both social structure and the genetics of intelligence must be considered. However, these complications do not alter the basic features of the simple model and they can be treated summarily. What we want to know of each complication is its effect on the rate of social class differentiation in $I_A$. We may therefore divide them according to whether they speed up or slow down the differentiation process.

Among the accelerating factors must be included a higher rate of social mobility and the differentiation of more social classes. It may be noted, parenthetically, that a uniform increase of mean $IA$ with social class need not be assumed. An interesting case of this kind is suggested by the recent British social mobility inquiry in which the Index of Association (i.e. the degree to which caste conditions are approximated) rises sharply at the highest occupational levels (Glass 1954, p. 199). If the high caste in our Case 1 were allowed to remain a caste while selective migration for $IA$ were allowed to produce a middle class out of the former low caste, then the middle class would inherit a mean $IA$ higher than that of the high caste, largely by recruitment from the heterozygote pool in the low class.

A more formidable array of decelerating factors comes into play as the model is modified to take account of real conditions. Of these the first is the assumption of polygenic determination of $IA$ – and the trend in psychology towards the recognition of more factors in measured human abilities implies a more complex genetic structure determining intelligence. In fact, it is difficult to conceive of anything as elaborate as human intelligence being determined by as few as (say) ten genes. Yet if we add only one other gene $B$ in Case 2, with all other assumptions, both genetic and social, remaining the same the result is that the high class does not become 'pure'[5] before the passage of nine generations.[6] This process is summarized in Table 11.2.

Second, on the side of social structure we have to recognize the existence of determinants of social mobility which are wholly unconnected or only partially connected with intelligence – for example, sexual beauty or ugliness, good or evil fortune in

Table 11.2 Social distribution of intelligence (IA) in a class society with two-gene inheritance of intelligence

| | Low class | | | | | | | High class | | | | | | |
| | Gene frequency | | Number of persons | | | | | Gene frequency | | Number of persons | | | | |
| Genera-tion | $pa^L$ and $pb^L$ | $p_A^L$ and $pB^L$ | aabb | aAbb or aaBb | AAbb or aaBB | AABb or aABB | AABB | $pa^H$ and $pb^H$ | $pA^H$ and $pB^H$ | aabb | aAbb or aabB | AAbb or aaBB | AABb or aABB | AABB |
| | (1) | (2) | (3) | (4) | (5) | (6) | (7) | (8) | (9) | (10) | (11) | (12) | (13) | (14) |
| $F_0$ | 0·500 | 0·500 | 56 | 225 | 338 | 225 | 56 | 0·500 | 0·500 | 6 | 25 | 38 | 25 | 6 |
| $F_1$ | 0·510 | 0·490 | 61 | 234 | 337 | 216 | 52 | 0·409 | 0·591 | 3 | 16 | 35 | 34 | 12 |
| $F_2$ | 0·519 | 0·481 | 65 | 242 | 337 | 208 | 48 | 0·327 | 0·673 | 1 | 9 | 29 | 40 | 21 |
| $F_3$ | 0·528 | 0·472 | 70 | 250 | 335 | 200 | 45 | 0·250 | 0·751 | 0 | 5 | 21 | 42 | 32 |
| $F_4$ | 0·535 | 0·465 | 74 | 256 | 334 | 194 | 42 | 0·186 | 0·814 | 0 | 2 | 14 | 40 | 44 |
| $F_5$ | 0·541 | 0·459 | 77 | 262 | 333 | 188 | 40 | 0·130 | 0·870 | 0 | 0 | 8 | 34 | 57 |
| $F_6$ | 0·547 | 0·454 | 80 | 266 | 332 | 184 | 38 | 0·082 | 0·918 | 0 | 0 | 3 | 25 | 71 |
| $F_7$ | 0·550 | 0·450 | 83 | 270 | 331 | 180 | 37 | 0·047 | 0·953 | 0 | 0 | 1 | 16 | 82 |
| $F_8$ | 0·553 | 0·447 | 84 | 273 | 330 | 177 | 36 | 0·019 | 0·981 | 0 | 0 | 0 | 7 | 93 |
| $F_9$ | 0·556 | 0·444 | 86 | 274 | 329 | 176 | 35 | 0·000 | 1·000 | 0 | 0 | 0 | 0 | 100 |

*Note*: For purposes of presentation the figures in columns 1, 2, 8 and 9 are rounded to three decimal places and in all other columns to whole numbers.

business, military prowess or timidity, etc. Though precise estimates are impossible, we can be confident that selection other than for $IA$ has been of great relative importance especially before the rise of the educational system as an agency of social selection in the last two generations. This type of social mobility or any increase in its incidence would have decelerating effects on the process of differentiation in the $IA$ composition of the classes. Moreover, any reduction in mobility based on selection for $IA$ would have a similar effect.

Putting the accelerating and decelerating factors together it seems reasonable to conclude that the net effect would be to slow down the process of differentiation described under the simple assumptions of Case 2. If this be so, then we are justified in concluding that what little we know of the history of social stratification in Britain would hardly yield an expectation of social class differences in innate intelligence of the order arrived at in Case 2.

*Differential fertility and the decline of national intelligence*
At this point in the discussion, and before turning to the social distribution of measured intelligence it is convenient to undertake a short digression and consider one further social variable, namely differential fertility. We do so because of the importance of this factor in connection with the long-standing controversy over the trend of national intelligence. In the cases we have considered, the mean innate intelligence of the total population $(I_A{}^T)$ is constant because random mating and equal fertility between classes (castes) is assumed. If, however, in these cases we substitute the assumption of higher net reproduction rates for the low class (caste) then the mean $I_A{}^T$ would decline.

A hypothesis that a trend of this kind has obtained in Britain since about 1870 remains tenable. The results of the Scottish Enquiry can be attributed to a temporary action of ameliorative environmental influences which must follow from the correlation of social status inversely with fertility and positively with measured intelligence $(IC)$ (Scottish Council for Research in Education 1949).

On the other hand, it is possible to produce a model of 'intellectual equilibrium' with appropriate genetic and social assumptions including that of an inverse relation between social status and fertility. Penrose's (1948) model is an example of this.

Moreover, what is known about social status, family size and intelligence indicates a complex set of relationships (see Anastasi 1956 for an excellent summary of the present state of knowledge). The inverse relation between status and fertility is neither simple nor universal. It is often reversed at the higher income and occupational levels and does not apply to the high fertility areas of the world (UN Dept. of Social Affairs 1953, Dinkel 1952). The negative correlation between family size and *IC* is well established where family size is defined as size of sibship but, so far, has not been found in studies where parental *IC* is correlated with number of offspring (Anastasi 1956, p. 195).

### 3 Measured intelligence (*IC*)

We may now return to our central problem. The existence of individual differences in a postulated *IA* remains as a 'legitimate hypothesis'. But especially if *IA* is polygenically determined, and given the history of social structure in Britain, their issue in social class differences is doubtful.

Further evidence in favour of this view may be had from an examination of the social distribution of measured intelligence (*IC*). The figures given in Table 11.3 relating to 1316 children entering secondary schools in south-west Hertfordshire in 1952 are are typical. For the total group of children they show a mean $IC^T$ of 100·97 with a standard deviation of 14·15. When the children are classified by paternal occupation the mean *IC* rises with occupational status and the variance in each occupational group is less than that of the total population.

Table 11.3 *The social distribution of measured intelligence*

| Father's occupation | Mean IQ | Standard deviation | Variance | Variance* |
|---|---|---|---|---|
| | (1) | (2) | (3) | (4) |
| Professional workers, business owners and managers | 112·95 | 11·62 | 135·01 | 67 |
| Clerical workers | 109·15 | 12·59 | 158·51 | 79 |
| Foremen, small shopkeepers, etc. | 103·70 | 13·47 | 181·44 | 91 |
| Skilled manual workers | 100·10 | 12·79 | 163·58 | 82 |
| Unskilled manual workers | 97·15 | 13·24 | 175·30 | 88 |
| Total | 100·97 | 14·15 | 200·22 | 100 |

\* Variance as percentage of total variance.
*Source:* Floud, Halsey and Martin (1957), p. 59.

Now the fraction of variation in *IC* which is attributable to heredity is a matter of dispute, but the estimates of $\frac{3}{4}$ and $\frac{1}{2}$ may be taken to represent the upper and lower limits (see Mather 1956). If the lower limit is accepted it follows that only where the variance in a subgroup is less than 50 per cent of that of the total population can we be certain that both heredity and environment are at work in the differentiation of the subgroup. Similarly, if the variance remains at more than 50 per cent, it is possible to entertain the hypothesis that the subgroup is differentiated solely on the basis of environmental circumstances. Thus it is clear from column 4 of Table 11.3 that, assuming the lower limit, the class differences in *IC* may be entirely environmental.

On the other hand, if the upper limit of $\frac{3}{4}$ is taken as the fraction of variation which is heritable the result still holds, with the single exception of the professional group, the variance of which is 67 per cent of the variance for the total group.

Of course this analysis, even leaving aside the exceptional case, does not demonstrate that the differences in mean *IC* between the classes *are* caused by environment; it only shows that they could be. An experiment designed to test the hypothesis that these differences are caused by the environmental components of social class is, in principle, simple. It would be necessary to take a random sample from newly born babies among (say) the unskilled workers, distribute them at random in the other classes and, during and after upbringing, compare the mean *IC* of each of the 'displaced' and 'control' groups. In practice, however, the difficulties of such an experiment are enormous. The available comparisons most approximate to it are to be found in those twin and foster children studies in which an assessment of the social or 'cultural' level of the home is taken explicitly into account. Among the nineteen pairs of identical twins reared apart, studied by Newman and his associates (Newman, Freeman and Holzinger 1937), the average difference between members of a pair was 8·2 IQ points. Among six pairs judged to have the greatest discrepancies from the point of view of social advantage in the home, the average difference was 12 points. For the group as a whole, IQ differences correlated 0·51 with judged discrepancies in social environment. In studies of foster children, correlations of IQ with assessments of the cultural level of the foster home vary from 0·25 to 0·52.

We can, however, be quite sure that these results underestimate the environmental effects of social class. The types of homes into which the twins in any one pair were placed rarely differed very much and the placement of children in foster homes tends to be selective. In evaluating the studies of foster children, Anastasi and Foley state that 'if the total range of American homes were covered, reaching down to the most deficient, then the observed effect of home environment upon intellectual development would probably be greater'.

## Conclusion

Accepting the hypothesis of an innate intelligence (*IA*) we have drawn up a simple model of society designed to show the relation between variations in social structure and the distribution of genotypes.

In the extreme case of a caste society each caste has to be seen genetically as a separate gene pool and therefore, provided there is random mating within castes and equal fertility between them, the distribution of *IA* is stable.

When a caste society is transformed into a class society the classes may still be seen as gene pools but with migration between them. The question now becomes one of determining the relation between social mobility and genic migration. In our model of the class society we assumed that the rate of social mobility was such as to replace the 10 per cent of the high class with the least intelligence with an equal number of the most intelligent members of the low class. Under these conditions and assuming innate intelligence to be determined by a single gene, it took seven generations to produce a 'pure' high class. Even so, it is important to notice that, with a high class constituting one-tenth of the total population, the effect of draining off a pure high-intelligence high class from the reservoir of the low class scarcely affected the mean *IA* of the latter and still left it with a heavy preponderance of the total number of available genes for high intelligence.

The model may be elaborated in order to approximate more closely to the real conditions of contemporary society. For example, its fertility assumptions may be modified in order to throw light on the question of trends in the national intelligence. The central focus of our attention has been on the rate of genetic purification in the high class. From this point of view the refinement of our assumptions has to be considered in terms of acceleration and deceleration of the process of purification; and what we know of the genetics of *IA* and the history of social mobility in Britain leads us to doubt the hypothesis of innate class differences in mean intelligence.

From this and from a consideration of the small reduction in the variance of measured intelligence which results from classifying children according to paternal occupation, it seems reasonable to conclude that the observed differences between social classes in measured intelligence are more likely to be explained by environmental rather than genetic factors. Accordingly the research problem becomes one of disentangling the environmental components of social class which are relevant for intellectual development.

## Notes

1 In preparing this paper I have had helpful discussions with E. Caspari, R. Savage, M. E. Beesley and Mrs J. E. Floud. Mrs E. Proschan has helped me with the statistical calculations.

2 In Vernon's lucid discussion he distinguishes Intelligence A, B and C. A is 'inherited', B is 'largely acquired' and C is test performance.

3 Suppose a population to consist of equal numbers of *MM* and *Mm* individuals. Then the relative frequency of *M* to *m* is 3 to 1. Now consider the mating of an *MM* with an *Mm* individual. The offspring could perchance all be *Mm*, thus reducing the relative frequency of *M*. Thus, owing to chance fluctuation, especially in small populations, gene frequencies may increase or decrease. This phenomenon is called genetic drift.

4 Genetic drift could not, of course, be ignored in a complete analysis of such small populations.

5 It should be noted that we are looking at the rate of genetic purification of the classes and not the translation of those rates into means and standard deviations of scores. The latter would involve us in the complication, mentioned earlier, of increasing standard deviations in each generation.

6 However, for the addition of each further gene there is a smaller increase in the number of generations required to reach purity.

# 12 In search of an explanation of social mobility

## M. Young and J. Gibson [1]

(From *British Journal of Statistical Psychology*, vol. 16, pt 1, May 1963, pp. 27–35.)

Studies in different countries, which have shown that rates of mobility are more alike than they were expected to be, have raised questions about the mechanisms underlying social mobility. The following paper puts forward the suggestion that social mobility belongs to a cybernetic mechanism whereby something like a 'steady state' is maintained in each occupational class by means of constant movement into and out of it. The quality discussed is 'intelligence'; and the argument stems from two main observations. The first is that the higher the occupational class, the higher is the measured intelligence. The second is that the children of parents towards the extreme show a 'regression' from that extreme towards the mean of the general population. If these two observations hold true over a period of time, it follows that there must be social mobility. In a third part of the paper it is argued that differential fertility between the classes gives rise to additional mobility. The last section discusses longer-term changes which may influence the amplitude of short-term oscillations between any two generations, and finally notes a certain parallelism with the laws of thermodynamics.

## 1 Approach to the subject

Sociological researchers have during the last decade made many measurements of rates of social mobility for different generations and in different countries. For comparisons between generations in one country the Glass (1954) report (see also Sorokin 1927) is still the best source. When ten-year age-cohorts born from before 1890 to after 1920 were compared, very little difference was found in rates of mobility. Roughly the same proportion in the younger

as in the older cohorts had occupational status higher than, or the
same as, their fathers: 'in general the picture of rather high
stability over time is confirmed'. More recently Lipset and Bendix
(1960) have, in a comprehensive review of the literature,
compared different countries. They measure the rate of social
mobility upwards by the proportion of the sons of manual workers
who become non-manual, downwards by the proportions of the
sons of non-manual workers who become manual. In urban areas
rather similar proportions of sons move up in a number of
countries – in France 35 per cent, in Germany 26–30 per cent, in
Switzerland 44 per cent, in Sweden 29 per cent, in Japan 33 per
cent, and in the United States 31–35 per cent. There is more
discrepancy in the proportions moving down, varying from 13 to
38 per cent. Studies in particular cities also show rather similar
rates in Poona and in Tokyo, in Sao Paulo and Kansas City, in
Aarhus and Indianapolis. A recent study of a London suburb,
Woodford, produced the same results (Willmott and Young
1960). If, instead of a two-class system, a division is made into
several more classes, the rate of mobility – say, into and out of the
professions – would probably be less strikingly similar (Miller
1960). But at any rate the general conclusion of the work done so
far is that rates of mobility, over time and space, are much more
alike than they were expected to be.

Now that some measurements have been made, however
imperfect, the emphasis in the next phase of research should be on
finding out what are the mechanisms underlying social mobility.
But obviously some ideas will be needed beforehand about what
these mechanisms might be. The purpose of this paper is to put
forward one such suggestion, which, if borne out, would perhaps
go some way towards explaining the similarity in mobility rates.
We propose to test this suggestion in field research. We are both
aware of the excesses of Social Darwinism in the nineteenth
century, and concerned to prevent a repetition of them in the
twentieth; but we are also keen to conduct hybrid research of
interest to a biologist as well as a sociologist. Our hope is that other
investigators in other places may be tempted into similar
interdisciplinary research.

*Cybernetic mechanism*
The suggestion is, in brief, that social mobility is the variable
element in what Cannon (1948) called a homeostatic and Wiener

(1948) a cybernetic mechanism, whereby something like a 'steady state' is maintained in each occupational class by means of constant movement into and out of it. The mechanism is in some ways similar to that envisaged in natural selection in the Malthusian scheme and in Ricardian price theory. We shall discuss how it might work in relation to the distribution, first of intelligence, and secondly of fertility, between occupational classes.

We are here defining 'intelligence' rather narrowly as the qualities measured by the admittedly highly fallible instrument of intelligence tests – orthodox ones, not of the type which tests 'creativity' (Getzels and Jackson 1962). This is not the only quality to take into account in any full investigation. The tenor of what Fisher said about what he called 'social promotion' rather than 'social mobility' is clearly right, even though the expression now sounds somewhat Victorian: 'What is perhaps more important is that a number of qualities of the moral character, such as the desire to do well, fortitude and persistence in overcoming difficulties, the manliness of a good leader, enterprise and imagination, qualities which seem essential for the progress, and even for the stable organization of society, must, at least equally with intelligence, have led to social promotion' (Fisher 1930). Such characteristics are important because they affect not only 'social promotion' but also the adaptability of a class as much as of a society. As Thoday (1963) has said, 'We cannot assess just how much genetic diversity is needed in our complex and ever-changing social systems, but we can be sure that it is a good deal'. Yet we have to start somewhere; and intelligence, although not the sole quality to consider, is at least an important one.

## 2 Mobility and intelligence

Our argument stems from two main observations. The first is that the higher the occupational class, the higher the measured intelligence (IQ for short). The usual figures of this sort, for instance those collected by Floud, Halsey and Martin (1956), are for children only. It has been shown again and again in many different countries that the children of (say) professional class parents score better than those of manual workers. There is less published information for adults: most intelligence tests are designed for children, not adults; but what there is, for instance, on the testing of recruits for the Armed Services (Conrad and Jones 1940), shows that there are the same sorts of differences for adults, although always more marked for the adults in a particular class than for the children. The figures in Table 12.1 illustrate this from a paper by Burt which marked an important advance (Burt 1961).

Table 12.1 *Mean IQs of parent and child according to class of parents*

|                      | Parent | Child |
|----------------------|--------|-------|
| Higher professional  | 139·7  | 120·8 |
| Lower professional   | 130·6  | 114·7 |
| Clerical             | 115·9  | 107·8 |
| Skilled              | 108·2  | 104·6 |
| Semi-skilled         | 97·8   | 98·9  |
| Unskilled            | 84·9   | 92·6  |
| Average              | 100·0  | 100·0 |

## Galton's Law

The second observation is that stated in Galton's 'law of regression' (Galton 1889) – namely, that the children of parents towards the extreme show a 'regression' from the extreme towards the mean of the general population (Anderson, Brown and Bowman 1952). Fathers (or mothers) with mean IQs of about 84, or 140, have children with IQs of about 92, or 120, respectively. This again is illustrated by Burt's table. Since the children from any particular class have a greater spread of intelligence than their parents, the shape of the distribution for the population in general is not progressively pinched, but remains roughly the same from generation to generation. Another way of putting it is to say that there is a correlation of less than unity between the IQs of parents and children. From Galton onwards many investigators have found correlations in the region of 0·5 between parent and child, and between sib and sib, not only for IQ, but also for other graded characteristics, such as height and span of arms.

Why regression? [2] Is it due to genetic influences, to the effects of mating, or to environmental influences? We shall say a word about each. First, the genetic variable: (see Fisher 1918 and Penrose 1949). The correlations of around 0·5 which have been found are about what would be expected theoretically if there were complete hereditary determination of IQ – but only on four assumptions, that several identical genes with additive effects are involved, that the gene pairs do not show dominance or recessiveness, that there is no assortative mating, and no environmental variations. The first two assumptions are difficult to test, though, as Burt and Howard (1956) have shown, there is good evidence that the genetic element is mainly multifactorial. But the last two assumptions are unrealistic. Mating is assortative; the correlation between husbands' and wives' IQs is about 0·5 (see Dahlberg 1948). The parent-child correlation should therefore be considerably above 0·5. It follows that the fourth assumption must also be unrealistic. The environments of parents and those of their own children are often similar, and this in part accounts for the similarity in intelligence; the fact that they are not identical accounts for much of the dissimilarity, and so tends to bring the correlation back to 0·5. Parents towards the extreme of the IQ distribution probably had exceptionally favourable or unfavourable environments and on chance grounds alone their environments are likely to be more exceptional than those of their children.

*Necessity for movement*
So we can say quite firmly that genetic, mating, *and* environmental influences on parent-child correlations are at work, even though in the present state of knowledge their respective effects may be difficult to sort out. Fortunately all that is necessary for our argument is the fact of regression. Given that, and given that the first observation we have referred to – the positive correlation between class and intelligence – is one that applies to different points of time, it follows that there must have been mobility upwards of more intelligent children to take the place of the less intelligent who move down from the higher classes (for a review of the somewhat meagre evidence, see Burt 1959). Had such mobility upwards and downwards not occurred, the mean IQs for the parents in the upper classes would not be as high as appears in Table 12.1, but would be at least as low as the means for the children, or indeed, if the regression had continued without compensation for a series of generations, they would by now have reverted to the general average; and the distribution of intelligence in each class would have become much more like the distribution in

the general population than it actually is. A rough stability can only be maintained within each class by a continuous interchange between them (see Conway 1958 and Halsey 1959).

In the absence of inquiry, we do not know what has been happening over time to the distribution of intelligence in the various classes. But we have made a small study of a kind not made before – an attempt to discover who, in IQ terms, actually moves up and down: this may serve to indicate the kind of material which could be gathered on a larger scale. A sample of forty-seven sons in their twenties was interviewed in 1962 in the town of Cambridge (students being excluded), and the forty-seven fathers in whatever part of England they were living. The results show that there was the same regression from fathers to sons as is found by all investigators, although less neat than that shown by Burt's figures (from a larger sample) in Table 12.1. But the effect of regression was offset by 'exchange' of sons between the classes in the way that our idea suggests should happen. Sons more intelligent than their fathers by and large moved up, and those less intelligent down; and, more important, if the distance of movement was measured on a six-point class scale and related to the extent of the difference between the intelligence of fathers and sons, the greater the difference in intelligence, the greater the distance of movement. The consequence of this reshuffling was that the correlation between intelligence and class for non-manual workers as a whole was very nearly restored in the sons' generation to the level it had been in the fathers'. The correlation which had been 0·70 for the fathers came back to 0·68, as the result of mobility, for the sons. The Cambridge study therefore supports the suggestion we are making in this section of the paper.

## 3 Mobility and fertility

We will now briefly consider the effects of differential fertility between classes. Its relevance is obvious. If, for instance, intelligent adults in the upper classes do not by and large completely reproduce themselves, there will need to be movement upwards of intelligent children to maintain a stable correlation between intelligence and class. The first part of this proposition is probably not unrealistic (see Burt 1952). It was at any rate accepted by the Royal Commission on Population: 'Summing up what we may regard as reasonably well established in the evidence about the differential birth rate, it can be said that, of the social groups, those with the highest incomes, and among individual parents within each social group, the better educated and more intelligent have smaller families on the average than others' (Royal Commission on Population 1949). Glass (1954) estimates how much mobility there would have to be to counter the effects of such differential fertility as was not fully offset by differential class mortality. Over the period with which he was concerned 300 non-manual fathers would have been replaced by only 252 adult sons, and 700 manual by 748 adult sons. To maintain stability, 5 per cent of sons would therefore have had to move up, and 5 per cent down, on this score alone.

Up until recent times it has been assumed that, if mobility was perhaps maintaining stability in the short run, it was only at the cost of disrupting it still further in the long run. Although not put in these terms, the idea was that the cybernetic mechanism was liable to a kind of increasing oscillation or 'hunting', with every correction creating a need for greater correction. This could have been the case on the ground advanced by Fisher: 'That the economic situation in all grades of modern societies is such as favours the social promotion of the less fertile' (Fisher 1930). For Fisher this was the reason why all civilizations have been eroded. For many others the same line of thought, buttressed by many studies which have shown a negative correlation between family size and intelligence, has suggested that, taking all classes together, intelligence must be declining (Anastasi 1956).

*Minnesota study*

This conclusion has, however, been called in question, both by comparisons of US servicemen in the two World Wars (Tuddenham 1948) and by comparisons over time of school-children in Scotland (Scottish Council for Research in Education 1949). These suggested no significant change in intelligence. More recently, Higgins, Reed and Reed (1962) in Minnesota have shown that much of the reasoning about intelligence decline has been questionable because it has not taken full account of childless people. They show that, although there is a negative correlation between intelligence of children and the number of their siblings, this does not result in any fall in intelligence because some of the most unintelligent children do not reproduce themselves. 'The higher reproductive rate of those in the lower IQ groups who are parents is offset by the large proportion of their siblings who never marry or who fail to reproduce when married.'

In any case the situation has been changed by the recovery in the birth rate in the higher classes which has occurred since the Hitler war in the United States, and (to a lesser but still marked extent) in Britain and some countries of the Commonwealth and western Europe. No one yet knows the reasons for this change. All that seems certain is that biological considerations will have to be taken into account along with economic and sociological evidence. From the standpoint of biology, what has happened could be partly explained by two different arguments. The first is that the process has been in some ways similar to what happens when there is artificial directional selection in other species for a specific trait. Fertility is reduced for a time at the extreme high side of the distribution of the population until eventually a recovery is brought about by the restoration of genetic balance. The same thing could have happened with human beings precisely as the result of the stress given to the single trait of intelligence since the Industrial Revolution. As a result, mating has probably been more assortative for intelligence than before. This could have precipitated a decline in fertility of the more intelligent followed by a recovery. The second argument allows for a crucial difference between man and other species – that conception can be, and is, limited voluntarily. As the use of contraception has spread downwards from the upper classes, family size has depended more and more upon willingness to have children. But it is possible that this quality of willingness is to some extent genetic. In that event one would expect that birth rates would fall, to begin with, in any class when it first took to contraception on a large scale, but would recover in time as the proportion of 'willing' people in the population increased. As Darlington (1960) said: 'Previously children had been born to parents merely in accordance with their ability to beget and bear them. Now they were born to parents in accordance with their willingness. Thus for the first time in evolution, parents who did not want children, or want so many of them, were selectively disfavoured. Conversely, a selection began which favoured specifically the property of wanting children'. One merit of this argument is that its strength could be tested by inquiry.

The upshot of this discussion is that differential fertility may not give rise to as much mobility in the future as it has in the past. If data similar to the Minnesota data are found elsewhere, the lower fertility of people who are very low in IQ, and probably in class too, may occasion downward mobility. At the other end of the distribution, the recovery of fertility amongst people high in IQ and class may be reducing the amount of upward mobility. But even though differential fertility is not the prompter it was, it is obviously important enough to be taken into consideration in any future inquiry.

## 4 The dynamics of mobility

The idea we have discussed would help to explain not only mobility in general but the somewhat surprising similarities between countries mentioned at the beginning of this paper. If the same cybernetic mechanism is at work in all societies, that would help to account for the likeness in rates. But the discussion has been cursory. Not only have we concentrated solely on one of the characters with a bearing on mobility, but we have not been by any means exhaustive about that. We have not, for example, considered the influence of the variance of intelligence within each occupational class: the greater the variance and the more the overlapping between classes, the more one would expect mobility to arise between them. Nor have we considered how the standards by which 'intelligence' is judged change from one generation to another: again, the more change there is in the criteria, the greater should be the mobility. Nor have we considered the influence of the change from one generation to another in the size of classes: if the classes at the top expand, mobility upwards must be increased.

We have for the sake of exposition presented the idea in terms of a simple homeostatic mechanism. This may be acceptable as far as it goes: there may be a tendency in the short run for return to a 'steady state' from generation to generation. But underlying the movement from one generation to another are almost certainly secular changes – the long-term changes which set the context for the short-term oscillation between any two generations. To be more specific, what we imagine is that before a society embarks on the process of industrialization, marked above all by the application of Weberian rationality (Weber 1947) or reasoning power to social affairs, the distribution of intelligence in any one class is not so dissimilar from that in the population at large. But as the process of industrialization and the application of science gets under way, intelligence is elevated beyond all other human characters and its distribution between the classes is pulled apart. The 'upper classes' gain more than their proportionate share of intelligence; and with every increase the rate of mobility required to maintain that share goes up. The 'equilibrium' or 'steady state' is, in other words, likely to be changing from one generation to another. But a limit or critical point may be reached, at which no further increase in mobility can be secured. This brings us back to the mobility studies already mentioned. Lipset and Bendix (1960) suggest that most of the industrial societies for which data exist may have reached such a critical upper limit. When and if mobility does knock up against such a limit there would be no further rise in intelligence in the upper classes, and those classes might as a result lose some of their previous buoyancy. The rate of economic expansion might then become less. Only comparative studies in different countries in various stages of economic growth could bring definition to such a view.

Before finishing, we should say that there is an alternative way of describing what we have in mind which has the merit that the starting point is not the notion of a 'steady state'. Substituting 'mental energy' for intelligence immediately suggests a parallelism with thermodynamics. There may, in other words, be some applications in society of the laws of thermodynamics. Energy more ordinarily conceived is, according to the First Law of Thermodynamics, conserved. There may be a similar tendency for mental energy to be conserved. We referred to this above when we mentioned 'genetic balance'. And just as energy is, according to the Second Law of Thermodynamics, subject to diffusion or entropy, so too may there be a tendency for mental energy to be diffused over all occupational classes. But if there is anything in the argument put forward in this paper, the entropic tendency in society is countered by constant interchange of mental energy between the classes. This interchange allows the quantity of 'mental energy' or 'intelligence' stored in the form of knowledge to be augmented from one generation to another.

This paper is a prologue to inquiry. What is implied is the nature of the inquiry that is required. It would have to follow the small Cambridge survey (referred to previously) in gathering information by means of interviews in two or more generations about IQs and occupations of parents and offspring, and about their education. One of the purposes would be to show the influence of education. It would also have to differ from the type of inquiry made so far into social mobility. None of the studies has been able to deal explicitly with differential fertility. Questions have been put, in one generation alone, to samples of sons, the sons being questioned about the occupations of fathers. Since the comparison has been between father–son pairs, each father in each class has been shown as having been replaced by just one son. If the influence of fertility is to be investigated, it will be necessary to start with a sample of people old enough to have had children and proceed to trace as many as possible of the children they actually have. It is this kind of inquiry which will, we hope, yield systematic comparative information on regression, fertility and mobility. It is time sociologists did more than just measure mobility; it is time they tried to examine the forces which underly it.

**Notes**

1 We should like to acknowledge the help received from Christopher Wallis with earlier drafts of this paper; from Sebastian and Mary Halliday and May Brenner with the small survey in Cambridge reported in the text; from the people who commented on drafts, including Dr Sidney Brenner, Sir Cyril Burt, Hubert Child, E. Cooper-Willis, Professor Ernest Gruenberg, Dr Frank Hahn, Dr A. H. Halsey, Sir Julian Huxley and Professor Thoday; and from the Joseph Rowntree Memorial Trust which met the costs of the Cambridge survey.

2 'Regression' is partly due to the inevitable scatter of measurements. Even when the same person takes the same test twice, the correlation is less than unity.

# PART FOUR

*I Q, Genetics and Race*

# 13 Race and mental ability

*Arthur R. Jensen*

(From F. J. Ebling (ed.), *Racial Variation in Man: Institute of Biology Symposium No. 22*. London, Blackwell, 1975.)

Races, both human and infrahuman, are now most generally viewed from a scientific standpoint as breeding populations which, though interfertile, are relatively isolated from one another reproductively, by geography, ecology, or culture, and which differ in the frequencies of various genes. These major subdivisions of a species, called races, are classifications based upon the relative degrees of intra-group similarities and inter-group differences in numerous genetically determined morphological, serological and biochemical characteristics. These genetic differences are products of the evolutionary process. Some of the many genetically conditioned characteristics in which various human races are known to differ are body size and proportions, hair form and distribution, head shape and facial features, cranial capacity and brain formation, blood groups, number of vertebrae, genitalia, bone density, fingerprints, basic metabolic rate, temperature, heat and cold tolerance, sweating, odour, consistency of ear wax, number of teeth, age of eruption of permanent teeth, fissural patterns on the surfaces of the teeth, length of gestation period, frequency of twins, male–female birth ratio, physical maturity at birth, infant development of alpha brain waves, colourblindness, visual and auditory acuity, ability to taste phenylthiocarbomide, intolerance of milk, galvanic skin response, chronic diseases, susceptibility to infectious diseases, and pigmentation of the skin, hair and eyes. Physical differences among some races are obviously extensive and profound.

There are also behavioural differences among races. In infrahuman species, behavioural differences among subspecies (i.e. races) are now generally viewed in an evolutionary sense as being continuous with the physical differences. Ethologists regard behavioural as well as physical traits as being subject to evolutionary change. An animal's behaviour can be a more important aspect of its adaptation to the environment than its physical characteristics, and can therefore play an important role in the evolution of the physical structures that mediate behaviour, principally the central nervous system.

The biological basis of behavioural differences among human races, on the other hand, has been much more in dispute. There has been the least consensus concerning the nature and causes of racial differences especially in those characteristics which most clearly distinguish *Homo sapiens* from all other species – a large, highly developed cerebrum and the capacity it affords for complex goal-conscious problem-solving behaviour involving planning, reasoning, judgement, imagination, decision – in short, intelligence.

My aim in this paper is to summarize as best I can from a scientific standpoint the main facts and theoretical issues involved in the study of human racial differences in behaviours commonly regarded as indicative of mental ability, without going into the background of sociopolitical and ideological controversy that continues to surround this topic. Readers should be told at the outset the three principles that mainly govern my own orientation in this inquiry.

First, I believe that objective research and objective knowledge are possible, and that it is desirable, indeed necessary, to guard the scientific aspects of the matter from entanglement with the political and social policy aspects. This is not to say that the latter are unimportant, but simply that we should strive as best we can to not let them in any way distort our aim of achieving an objective understanding of racial differences in mental abilities, limited though it may be, considering the intrinsic scientific difficulties.

Second, I emphasize the generally accepted position in science that explanations of phenomena are weak and unsatisfactory to the extent that they are *ad hoc*, and are more satisfactory to the extent that they are predicted by a more general theory or are consistent with some larger pattern of established systematic knowledge. That theory is best which yields the greatest number of verifiable predictions and the discovery of new phenomena, or can comprehend existing phenomena which previously had only *ad hoc* explanations. Evolutionary theory, population genetics, the polygenic theory of intelligence, developmental psychology, and psychometrics seem to me to provide the most comprehensive framework for the scientific study of population differences in abilities. It is my belief that explanations of racial differences which do not build upon the theoretical structures of these fields and their associated methodologies are the most likely to be invalid or scientifically unproductive. Whatever theoretical or methodological shortcomings these fields may have at present for the study of racial differences, I know of no better basis for formulating hypotheses and launching investigations.

Third, I believe we must accept the necessarily statistical and probabalistic nature of the evidence and conclusions in many aspects of this research. In comparing populations on psychological traits, we are, of course, dealing with continuous variation involving differences among frequency distributions with marked overlap. Distributions may differ in means, variances, skewness, and higher moments, and each kind of difference or combination of differences ultimately calls for theoretical explanation and has somewhat different implications. In all of the psychological traits we know of, it is frequently pointed out, variance among population means is much less than variance among individuals *within* populations. Moreover, since the causes of population differences in psychological traits are complex, involving many factors which cannot be experimentally controlled in research with human populations, our approach must be largely statistical. Rigorous proof of hypotheses, in the sense of logical necessity or the clear-cut ruling out of all alternative hypotheses by experimental control of variables, is not reasonably expected regarding most of the questions of greatest interest. We must make do, at least for the present, with conclusions expressed in terms of probabilities, often rather subjective probabilities at that, based on consistencies among converging lines of evidence and the weight that accrues to hypotheses by virtue of their integration with a larger theoretical framework, as opposed to *ad hoc* explanations. The tentative nature of conclusions at the growing edge of knowledge should always be kept in mind.

## Evolutionary differentiation

From the viewpoint of evolutionary theory, it is extremely improbable that any genetically conditioned characteristics, phy-

sical or behavioural, would have identical distributions of genotypes in all human populations. And the greater the evolutionary
separation between any two populations, the greater is the
probability of genetic differences in a wide variety of characters.
Geographical and cultural isolation of populations over many
generations results in cumulative differences in gene pools. The
specific evolutionary processes involved in the genetic differentiation
of populations are: gene mutations, random genetic drift,
selective migration, and natural selection.

*Mutation and drift*

Mutation and genetic drift are random processes occurring at
single gene loci, and consequently they are not major causal
factors in population differences in polygenic traits, i.e. continuous
traits, like height and intelligence, which are determined by a
large number of genes. The larger the number of genes involved
in a given trait, the less is the probability that random changes, or
drift, occurring at individual loci would all happen to act in the
same direction to produce large differences between populations.

The theory of genetic drift, however, permits calculations
concerning the relative degree of genetic isolation between
populations, based on the number of differences that would occur
by random genetic drift alone, without considering the greater,
systematic and directional differences brought about by selection.
On this basis, for example, geneticists have estimated the
'divergence times' or extent of genetic separation between the
three major races as about 14,000 years between Caucasoid and
Mongoloid, 42,000 years between Mongoloid and Negroid, and
46,000 years between Caucasoid and Negroid (Nei and Roychoudhury 1973). These estimates were based on the observed
differences in the frequencies of neutral genes, i.e. genes for which
there is no evidence of selection. The divergence time is the time
that genetic drift by itself would take to make the frequencies of
neutral genes differ between the major races as much as they do at
present. This means, in other words, that these three major racial
groups have been separated long enough and completely enough to
permit a purely random genetic drift in gene frequencies
equivalent to some 2000 generations of complete separation
between the Negroid and the other two races, and about 700
generations of complete separation between the Caucasoid and the
Mongoloid. However, it should be remembered that these

differences due to drift would be expected to have only minor explanatory significance for racial differences in polygenic traits, especially traits which have been subject to natural selection.

It is now possible rigorously to measure the evolutionary distances between various species, as has been done with chimpanzees and gorillas, in terms of the degree of similarity of DNA sequence in certain blood proteins, and to measure the evolutionary relatedness of man to the other primates. But as yet this method is not sufficiently developed to delineate the evolutionary distances among human races with any reasonable precision.

## Migration

Migration *per se* is probably not a major factor in producing population differences in polygenic traits. But migration often involves selection, either of the original migrant population or of subsequent generations, since having to cope with the challenges of an alien environment affords new opportunities for selection to alter the gene pool of the migratory groups. For example, migration from a tropical to a temperate clime could involve selection of whatever genes might be involved in the capacity for the planning and foresight needed to survive the long winters (we know from experimental behaviour genetics with animals that the capacity for acquiring almost every behavioural characteristic, including the general capacity for learning, responds to selection). Also, plagues and famines which often accompanied migrations produced genetic 'bottlenecks' in human populations. That is to say, a relatively large population would be reduced for a few generations to a small, highly selected breeding group, with statistically different gene frequencies from those of the parent population, which then grows again into a large population. Such 'bottlenecks' can result in marked changes in the gene pool within a relatively short period, depending upon the nature and severity of the selection.

## Selection

Natural selection is by far the most probable evolutionary mechanism causing the major differences between human races, especially as regards polygenic traits. When a complex phenotypic characteristic, physical or behavioural, involves the influence of a number of genes, all the genes are selected simultaneously, since selection acts directly on the phenotypes. The rapidity of selection for the relevant genes depends both upon the severity of the selection pressure on the phenotypes and upon the narrow heritability of the characteristic in question, i.e. the proportion of phenotypic variance attributable to additive genotypic variation. Selection, so to speak, tends to use up the additive genetic variance, since it is that part of the individual's genome which is most highly correlated with the phenotype. As selection proceeds, the narrow heritability of the trait decreases; that is, there is less additive genetic variance and an increasing proportion of the genetic variance is attributable to dominance deviation (i.e. interaction or non-additive effects of alleles at the same loci) and to epistasis (i.e. non-additive effects of genes at different loci). The presence of non-additive genetic variation, which can be estimated by the methods of quantitative genetics, therefore, indicates that the trait in question has undergone selection, and if the dominance is for either high or low values of the trait, it means there has been directional selection.

It is highly significant to our inquiry, therefore, that the appropriate quantitative genetic analyses of scores on standard tests of intelligence show some dominance and other non-additive genetic variance, as much as 10 to 15 per cent (Jinks and Fulker 1970, Jinks and Eaves 1974). Dominance for high IQ indicates there has been directional selection. Thus, whatever ability is measured by IQ tests, it shows the 'genetic architecture' expected for a fitness character; the IQ apparently reflects some trait of biological relevance in human evolution.

How might this have come about? Cranial capacity, a crude measure of intelligence, is known to have increased markedly over the 5 million years of human evolution, almost tripling in size from the earliest fossil information of *Australopithecus* to present-day man. The greatest development of the brain was of the neocortex, especially those areas serving speech and manipulation. Tools found with fossil remains indicate that increasing brain size was accompanied by the increasing complexity of tools, and along with the development of complex tools are also found artistic drawings on the walls of caves. In the last 1 or 2 million years the strongest selection pressure in man has been for behavioural traits of increasing complexity, accompanied by the increasing size and complexity of the cerebrum. The ethologist Konrad Lorenz (1973) has elaborated upon the thesis that the evolution of the complex functions of the human brain that make possible such intelligent operations as comparing, analysing, separating, seeing relationships, classifying, counting, abstracting, conceptualizing, recalling, imagining, planning, and the like, came about from selection by environmental demands acting directly upon the behaviours made possible by increasingly complex nervous functions.

It seems highly probable that such powerful selective processes have also operated to some extent differentially upon subgroups within one species that have been genetically isolated for thousands of generations. Therefore, in terms of evolutionary theory, behavioural traits and their genetic and physical underpinnings in the nervous system should be expected, with high probability, to differ among human races. If our psychological measurements did not reflect such differences, they would seem highly suspect, since, in principle, differences are practically certain to exist.

We can only speculate about which specific selection mechanisms were probably most importantly involved in evolutionary differences in the behavioural capacities now called cognitive ability or intelligence.

Perhaps the most important general factor in selection for brain size and complexity was the presence of other men, making for competition for the means of survival, and selection for increasing ability to cooperate in hunting animals and in conflicts with hostile tribes. The invention of new tools and weapons and the development of skill in their use by other individuals would have conferred differential advantages making for selection. Each new invention in a sense divides the population into those who can and those who cannot learn to master its use, and tends to select in favour of those who can.

Population size is an important factor in the selective advantage of invention. The larger the group, the greater is the number of exceptional individuals most likely to make discoveries and inventions. New inventions and novel variations of existing tools and their correlated skills are less likely to arise in the relatively small and culturally isolated groups characteristic of primitive societies. Moreover, when an innovation does occur, and especially if it is a great advance beyond existing knowledge or skill, it may not be perpetuated unless some reasonably substantial number of the group can take it up. Depending upon its degree of novelty and complexity, they would have to be the more exceptionally able individuals, and, given the normal distribution of abilities, more such able individuals would exist in a larger population, so that a new invention of only one exceptional member of the group would take on selective significance for some substantial number of the population.

Inventions and discoveries involving tools, weapons, skills, and knowledge about the environment of adaptive importance, create greater salience of individual differences in abilities which then become important factors in selective and assortative mating. As one moves from relatively primitive to relatively advanced societies, individual differences in cognitive ability become more conspicuous and more consequential in many ways that can affect an individual's fitness in the Darwinian sense. In a number of early human societies mating was a prerogative of the ablest and most esteemed males, each of whom had many females, while many less esteemed males had no mates.

Evolutionary rates for certain traits could differ considerably among groups with different mating customs or different degrees of selective mating for various traits. In considering natural selection for abilities in man, one must consider what proportion of a population is regarded by its members as subnormal or in any way undesirable from the standpoint of selective mating, and this will, of course, depend to a considerable extent upon the nature and cognitive complexity of the cultural demands made by the society. Even a slight reproductive advantage can have marked genetic consequences on the time scale of human evolution. For example, it can be calculated that a gene that confers a 1 per cent reproductive advantage in a population will increase in frequency from 0·01 to over 0·99 in 1000 generations, assuming that the same degree of advantage is maintained throughout this period.

Increased population size also decreases the degree of inbreeding and gives rise to more new genetic combinations which are grist for selection.

Primitive societies consisted of hunter-gatherers, and for obvious ecological reasons were kept relatively small in numbers. The advent of agriculture permitted population densities a thousand times greater than those of hunter-gatherers, thus magnifying the selection factors for cognitive abilities associated with a larger population. Also, in terms of abilities for counting, measuring, planning, mastering the environment, and a greater complexity of social, political, and economic organizations, agriculture probably placed a higher premium on intelligence than did hunting and gathering. In fact, civilizations grew up along with the development of agriculture. Various populations of the world differ in thousands of years in the time since they abandoned hunting and gathering for agriculture, and some presently existing groups have never taken up agriculture.

Thus, in general terms, man's evolutionary history and the relative isolation of various populations for thousands of generations would justify the expectation of genetic differences between populations in a host of characteristics including those in which selection pressures have acted differentially upon behaviour. These behaviours would be mainly polygenic traits for which population differences are statistical rather than typological. It would seem most improbable that at least some of the genetically conditioned behavioural differences that have come about in the course of evolution would not be among the observable differences between contemporary races. A contrary view would have to argue one of four propositions: (1) the selection pressures in all long-term isolated populations in the course of human evolution have been identical for all groups for all abilities; (2) even if there have been different selection pressures for different components of ability, these components would average out to the same value in their combined effects on performance in every population, provided there is equality of opportunity for the development and expression of abilities; (3) there is only one general ability that is inherited – a highly plastic capacity for cultural learning which is genetically the same in all populations and becomes differentiated only through environmental and cultural influences; or (4) even if there are genetic ability differences between populations, they are so obscured by cultural and environmental factors that there is zero correlation (or even a negative correlation) between the distributions of phenotypes and genotypes. Numbers 1 and 2 have the disadvantage of being extremely improbable. Number 3 is contradicted by the factor analytic and behaviour-genetic analysis of mental abilities, which reveal a number of different abilities with relatively independent genetic bases. The fourth point seems more debatable, since it depends so much upon the methods of measuring abilities and the extent of the cultural differences between the groups in question. Modern students of racial differences have seemed most reluctant to point to various aspects of particular cultures as being in themselves indicative of

differences in mental abilities. However, John Baker, an eminent
biologist who has written recently on the subject of race, notes the
fact that racial groups have differed quite markedly in the degree
to which they have developed 'civilization' (in terms of a list of
twenty-one criteria ordinarily regarded as indicative of being
civilized) and also the degree to which complex cognitive abilities
are manifested or demanded in various societies. The Arunta
language of Australian aborigines, for example, conveys only the
concrete; abstract concepts are not represented, nor is there any
verbal means of numeration beyond 'one' or 'two' (Baker 1974,
pp. 500–1). Baker notes that these various criteria of cultural and
intellectual advancement rank order existing races much as do
standard tests of cognitive ability when applied to representative
members of these racial groups who have been reared under
similar conditions of civilized life. Baker's book is replete with
specific factual examples and comparisons of racial groups in
terms of these various criteria. He concludes: 'the reader will not
have overlooked the fact that repeatedly, in each relevant context,
the possibility of environmental causes has been reviewed in some
detail and rejected as an insufficient explanation of the facts'
(p. 533).

## The heritability of group differences

The polygenic theory gives a quite good account of *individual*
variation in intelligence. In principle, at least, it is also applicable
to *group* differences in ability, which are viewed as qualitatively
the same as individual differences. The gene pools of relatively
isolated populations are hypothesized to differ in the frequencies
of the genes involved in abilities. But the relevant genes are the
same in all populations, so their differences are quantitative, not
qualitative. The polygenic theory itself is completely agnostic as to
the direction and magnitude of the genetic difference between any
two specific populations. In this respect the polygenic theory
contrasts markedly with the environmentalist view, which maintains that there are no genetic differences in mental abilities, or at
least in general intelligence, among any human populations.

It may seem surprising that, in practice, the polygenic theory
yields few predictions concerning differences between particular
races which are testable by means of any presently available
evidence. One type of prediction concerns the intelligence of racial
hybrids. The polygenic theory predicts the mean scores (say, IQ)

of the hybrid offspring to be approximately intermediate between the means of the two different racial populations. The only studies of this type reported in the literature are of white and Negro crosses. These studies are generally unsatisfactory, as I have pointed out in detail elsewhere (Jensen 1973a, pp. 219–30), since there is reason to believe that persons entering into interracial marriages are probably not representative of their populations in intelligence. Most studies of the intelligence of racial hybrids are not based on known pedigrees, but on the selection of hybrid subjects solely on the basis of their physical appearance being more or less intermediate between Caucasoid and Negroid in such characteristics as skin colour, nasal width, and interpupillary distance. The majority of such studies are in accord with the genetic prediction, i.e. the intermediate group in appearance also usually stands between the more 'pure' appearing racial groups in mental test scores. Also, in twelve out of eighteen studies of American Negroes with some Caucasian admixture, there was a significant positive correlation between skin colour (lightness) and IQ. Although these studies leave little doubt of a relationship between skin colour (and other racial characteristics) and IQ, they are a weak test of the genetic theory, since the same correlation could result from cross assortative mating for skin colour and IQ within the Negro population without any necessary implications concerning the direction or magnitude of a possible genetic difference in IQ between the Negro and white populations.

To overcome this problem, it has been proposed to use socially invisible genetic polymorphisms, which differ in known frequencies in West Africans and Europeans. These blood polymorphisms could be used as an index of racial admixture which would be correlated with IQ independently of visible racial characteristics such as skin colour. Aside from the technical difficulties in such research, which I imagine are surmountable, there seems a serious conceptual problem with this approach due to the fact that little is known about the selection that entered into interracial matings during the period of the greatest gene flow from the white to the American Negro gene pool, which occurred during slavery. The Negro population of the United States now has an average Caucasian admixture of about 25 per cent. We do not know how representative of the white population in intelligence were those individuals (practically all males) who practised miscegenation. If they were predominantly from the lower half of the white

IQ distribution and their mates predominantly from the upper half of
the Negro IQ distribution, the genetic consequences of hybri-
dization on the IQ distribution of subsequent generations of
American Negroes could be negligible or undetectable by any
presently available methods of genetic analysis.

Another prediction from genetic theory involves the phenomenon of 'regression to the mean'. The offspring of exceptional
parents, i.e. those who deviate above or below the population mean,
average some value more or less intermediate between the
parental value and the mean of the population. Regression is also
observed in the case of siblings. Sibling regression is less likely to
be contaminated by environmental effects than parent–child
regression, since not infrequently a parent has grown up in a
quite different environment than is afforded to his or her own
children. Siblings reared together generally share a more common environment. The regression is strictly predictable from the
polygenic model, but the degree to which the empirical findings
approximate the prediction depends upon the heritability, $h^2$, of
the phenotypic measurements. The complement of the heritability, $1 - h^2$, consisting of environmental and error variance, can
be regarded as 'noise' obscuring the prediction. For traits of high
heritability, such as height and intelligence, the predictions are
confirmed fairly precisely. Since the theoretical genetic correlation
between siblings is 0·5 under random mating and slightly
higher (about 0·55) under the degree of assortative mating
generally found for intelligence, one should expect, on average,
that the IQ of a given child's sibling would be just about half-
way between the given child and the population mean. Thus, it
is predicted from the genetic model, for example, that the
siblings of white and Negro children who are perfectly matched
for some given IQ would, on the average, have different IQs,
since the Negro sibs regress toward the Negro population mean
and the white sibs regress toward the white population mean. If
the two populations differ by about one standard deviation (or
fifteen IQ points) the two groups of siblings of the IQ-matched
Negro and white groups should differ half a standard deviation.
If the two IQ-matched racial groups both have an average IQ of
120, for example, the average IQ of the Negro sibs will be
$\frac{1}{2}(120 - 85) + 85 = 102\cdot5$ and the IQ of the white sibs will be
$\frac{1}{2}(120 - 100) + 100 = 110$.

This prediction was borne out in a study of all the Negro and white siblings in the elementary schools (ages 5 to 12) in a California school district (Jensen 1973a, pp. 117–19). The siblings of both white and Negro children were found to regress a constant fraction, about one-half, to their respective population means and not to the mean of the combined populations. This holds throughout the IQ range from about 50 to 150; the regression line, for both Negroes and whites, is linear throughout that range. Thus, this is a successful prediction from the genetic model. But it cannot be regarded as a proof of a genetic difference between the two populations, since the lower population mean of the Negro group, it could be claimed, is a result of a uniform environmental disadvantage or test bias in the Negro population. Thus, all that the sibling regression demonstrates rigorously is that the correlation between sibs is about the same in the white and Negro populations. A strictly environmental explanation of the mean population difference is not ruled out by this evidence. But an environmental explanation of it is *ad hoc*, unlike the genetic explanation, which is derivable from a pre-existing polygenic model. The polygenic theory would be in serious trouble if the prediction were not borne out. But there is no environmental theory that would have predicted the quantitative aspects of these results or the linearity of sibling regression throughout the normal range of IQs. In an *ad hoc* environmental account of the results, it would have to be regarded as a remarkable coincidence that environmental factors would so closely produce the same quantitative effects as are predicted by the genetic model.

Essentially, the reason that the regression phenomenon by itself does not prove genetic difference between populations is that even if one grants the same degree of heritability of a trait *within* each of two populations, and even if the heritability is very high, it cannot be inferred with certainty that the difference *between* the populations has a genetic component. It could be all environmental, or all genetic, or anything in between.

It is generally agreed that heritability within groups, $h^2_W$, has no logically necessary implication for heritability between groups, $h^2_B$. This does not imply, however, that there may not be probabalistic implications of $h^2_W$ for $h^2_B$ or that there is no theoretical connection whatsoever between $h^2_W$ and $h^2_B$, given knowledge of certain other parameters.

Generally, for highly heritable characteristics within groups, phenotypic mean differences between groups also show a heritable component, even when there are obvious environmental differences between the groups. Often there is a positive correlation between genotypes and the environmental factors most relevant to the characteristic, e.g. skin pigmentation and amount of exposure to ultraviolet radiation.

Instances are rare where the direction of genotypic means is the opposite to that of the phenotypic means; more often phenotypic and genotypic means are positively correlated. If within-group heritability is high (i.e. greater than 0·5), one must hypothesize a larger environmental difference than a genetic difference to explain a phenotypic difference between group means, unless one also posits an additional hypothesis that the mean difference between groups is due to environmental factors which are not the same as those responsible for environmental variance *within* the groups.

A reasonable presumption (though certainly not proof) of genetic group differences seems to be related to the magnitude of the group difference and the heritability of the trait in question, as seen in the fact that few persons believe that the average difference in stature between Pygmies and Watusis is not largely genetic, despite their very different habitats, diets, and customs. The fact that the group mean difference is large (relative to the standard deviation within groups) and involves a trait of very high heritability, makes it seem reasonable to believe that the group difference is largely genetic (I know of no other evidence that it is genetic). The same kind of 'reasonable hypothesis' must also apply to other characteristics, including behavioural traits, in which there are substantial phenotypic differences and substantial heritability within groups, although, of course, the degree of plausibility will depend upon the magnitudes of the group difference and of the within-groups heritability of the trait in question, as well as upon other factors such as the nature and extent of environmental differences, if these are known, and whether their causal relationship to the trait in question is established.

## Formulation of between-groups heritability as a function of within-groups heritability

The geneticist Jay L. Lush (1968) proposed the following formula of the relationship of between-groups heritability, $h^2_B$ (i.e. the genetic fraction of the variance among the phenotypic group means) and the heritability in the whole population (i.e. the combined groups):

$$h^2_B = h^2 \left[ \frac{1 + (n - 1)r}{1 + (n - 1)t} \right] \qquad (1)$$

where $h^2$ is the narrow heritability in the whole population, $n$ is the sample size, $r$ is the intraclass correlation among the genic values (for the particular character in question) of members of the same group, $t$ is the intraclass correlation among the phenotypic values of the same group.

When $n$ becomes large,

$$h^2_B \cong h^2 \, (r/t). \qquad (2)$$

The heritability within groups, $h_W^2$, can be expressed as:

$$h^2_W = h^2 \, \frac{(1 - r)}{(1 - t)}. \qquad (3)$$

From equations (2) and (3), the geneticist De Fries (1972) derived the following formula for the heritability between groups:

$$h^2_B \simeq \frac{(1 - t)r}{(1 - r)t}. \qquad (4)$$

If there is a positive correlation between heredity and environment, this expression underestimates the heritability of the group difference. If the correlation between heredity and environment is negative, $h^2_B$ is overestimated by the formula. The relationship of between-group to within-group heritability for two groups with equal variance, normal distributions of the trait, and a mean difference of one standard deviation, can be shown graphically as in Figure 13.1.

The formula is obviously only of theoretical interest, since we lack information on one of the parameters, $r$, the intraclass genic correlation for the trait in question. Thus the formula gets us nowhere, unless, of course, one wishes to speculate concerning the probable value of $r$. But this is the very point in question. If the groups do not differ at all genetically, $r$ will be zero and $h^2_B$ will be zero. For groups whose means differ by one standard deviation, the *phenotypic* intraclass correlation, $t$, is 0·20. (The intraclass correlation $t = 0.20$ is most easily obtained from a one-way analysis of variance which partitions the total variance (say, of IQ) between groups and within groups. If the group means differ by fifteen IQ points and the $\sigma$ within each group is fifteen IQ points, then the between-groups variance $\sigma_B^2$ will be $(15/2)^2 = 56.25$, and the within-groups variance $\sigma_W^2$ will be $15^2 = 225$. The intraclass correlation is $t = \sigma_B^2/(\sigma_B^2 + \sigma_W^2) = 56.25/(225 + 56.25) = 0.20$.)

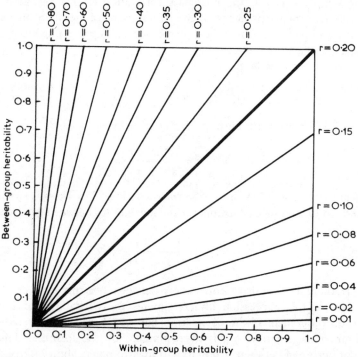

*Figure 13.1* Between-group heritability expressed as a function of within-group heritability and the genetic correlation of members of the same group ($r$). (From McClearn and De Fries 1973, p. 300.)

The genetic intraclass correlation $r$, however, is unknown. Unless one assumes that all the genetic difference between groups in the trait of interest is purely a result of random genetic drift (which affects all gene loci equally, on the average), there is no way I know of to estimate $r$ for any particular polygenic trait. And the traits in which we are most interested psychologically probably do not involve exclusively neutral genes. If they did involve only neutral genes and the trait were highly polygenic, then there would be no reason to expect any appreciable systematic genetic difference between large population groups. The size of $r$ will, of course, differ for various traits which have been subjected to different selection pressures over many generations. Thus it is pointless to try to estimate $r$ for one characteristic and expect it to be generalizable to others. De Fries (1972, pp. 10–11) states that 'Unfortunately, no valid estimate of $r$ is available'. But then he goes on to suggest a value of $r$ based on a coefficient of inbreeding. The value of $r$ is approximately twice the coefficient of inbreeding. He uses a coefficient of inbreeding estimated from morbidity data in Hawaii to arrive at a value of $r$ of 0·002. This is the average intraclass correlation (among different racial groups in Hawaii) for a random sample of all gene loci. There are many genes, perhaps the vast majority, that have not been subject to selection and which have similar frequencies in all human populations. Gene frequencies would differ only from random drift for most of the genes that enter into a coefficient of inbreeding estimated from morbidity statistics. Such an average over all loci does not provide any clue as to the intraclass genetic correlation for polygenic traits that have been subject to selection pressures as intelligence undoubtedly has. The intraclass genetic correlation for skin colour in Europeans and Africans, for example, would be much higher than 0·002 and probably approaches 1·00. The same would be true of height in Pygmies and Watusis. To what extent this is true for intelligence, we do not know. Obviously, the De Fries formula can yield no estimate of $h^2_B$ unless we can obtain an estimate of $r$ for the specific polygenic trait in question. An estimate of $r$ based on the average correlation over all loci, or on a random sample of genes, or on some other trait, simply will not do, and to base speculations on such estimates can only be misleading.

The De Fries formula, therefore, is useless empirically. Those who believe there are no genetic differences will say $r = 0$. If one makes the unwarranted assumption that genetic group differences are not confounded with environmental differences, then it might be said that $r = t/h^2$ (where $h^2$ is the heritability in the whole population). And if one makes the assumption that the between-groups environmental effects are of the same nature as within-groups environmental effects, one could say that $r = h^2 t$. But without making that assumption, which is crucial to the whole argument, we cannot know $h^2$ in the combined populations either, since this $h^2$ is a function of $h^2_W$ and $h^2_B$, and it is $h^2_B$ that we cannot determine for lack of knowing $r$. Because of this lack, we must conclude that, at present, attempts to infer the magnitude of heritability between groups is a blind alley.

## Within- and between-groups environmental variance

A knowledge of the heritability of intelligence within each racial group places some constraints on the magnitude of the mean difference between groups that can be accounted for in terms of all the environmental factors that contribute to variance within groups. The argument can be expressed most clearly in a series of points, using the well-established white–Negro IQ difference (in the United States) of one standard deviation, as an example.

(1) If the heritability $h^2$ of IQ is 0·7 to 0·8 in the white population (which is the best estimate we have from consideration of the total evidence), then the proportion of IQ variance attributable to environmental factors is $1 - h^2$ or 0·2 to 0·3. The standard deviation of the total environmental component of IQ thus can be calculated to be about six to eight IQ points (i.e. if $\sigma$ of IQ is 15, the variance is $15^2$ and the proportion of environmental variance would be $0·2 \times 15^2 = 45$, so the standard deviation would be $\sqrt{45} = 6·7$).

(2) If one assumes similar heritability of IQ in the Negro population, the standard deviation of the environmental component of IQ is about the same as in the white (i.e. item 1 above). (The evidence for IQ heritability in Negro populations, though not strong, does not suggest that $h^2$ differs appreciably from the estimate in white populations.) The existing correlations for twins and for siblings are highly similar for Negroes and whites. This does not prove that the heritability is the same in both groups, but it makes it the most likely hypothesis.

(3) If white and Negro populations differ, on average, by some fifteen to twenty IQ units, as the preponderance of the evidence indicates, then, given points 1 and 2 above, if it is hypothesized that all of this difference is environmental, it must be concluded that the groups differ by about two or three standard deviations in all the non-genetic sources of variance that make for IQ differences *within* the groups. Few would claim that the micro-environmental factors that constitute the *within*-families variance (e.g. birth order) should be included among the causes of the average difference between populations. It is the sources of

*between*-families variance, i.e. the kinds of environmental factors affecting all members of a family, that contribute to social class and racial group differences in IQ. The between-families environmental variance is about one-half to two-thirds of the total environmental variance within racial groups. This means that Negroes and whites, on average, must differ by some three to four standard deviations in such environmental influences if the standard fifteen to twenty points IQ difference is to be explained entirely in these terms.

(4) A variety of socioeconomic indices, singly and in combination, indicate that the average white–Negro differences in this respect is about one standard deviation or less – far from the three or four standard deviations of environmental difference that must be postulated by a strictly environmental hypothesis of the white–Negro IQ difference. In terms of these measurable (and potentially manipulable) kinds of environmental factors, studies of adopted children suggest that moving one standard deviation up or down on the environmental scale pushes the child's IQ up or down some six or seven points. Hence these kinds of environmental factors can account for only about one-third to one-half of the white–Negro IQ difference.

In the face of this analysis environmentalists hypothesize the existence of as yet unidentified and unmeasured factors, which produce IQ differences *between* racial groups but do not contribute appreciably to IQ variance *within* groups. Since no one has clearly specified the nature of these factors, I shall label them 'factor X'. 'Factor X' is purely *ad hoc,* invoked to explain the IQ gap still left when known measurable environmental differences are taken into account. Notions such as 'racial alienation', 'white racism', consciousness of being a minority, identification with a historically mistreated minority, etc., are attempts to characterize factor X. While these factors may exist it has not been shown, independently of the particular racial difference which they are invoked to explain, that they have any effect on IQ. And one may wonder why they do not apply to other minorities, such as Jews and orientals, who also have been subjected to discrimination, etc., but who score at or above the national average on standard tests, or to American Indians, whose environmental deprivations are the most severe of any subgroup in the US but whose performance on tests of mental ability and scholastic achievement is more or less intermediate as compared to whites and Negroes.

## Psychometric evidence

Although the discussion of racial differences from the standpoint of evolutionary theory and in terms of abstract principles of biometrical genetics can be carried on in general terms without reference to any particular racial groups, when we are faced with the prospect of actually making measurements and testing hypotheses we must get down to specific cases. At this point, understandably, there is often resistance or reluctance to our proceeding further. What may seem reasonable and intellectually acceptable in the abstract may seem odious and emotionally unacceptable when it comes down to specific cases.

It is a fact that the study of racial differences in mental abilities has focused much more extensively on sub-Saharan Africans and persons of African descent than on any other groups. Bibliographies of research on other racial groups are extremely scant by comparison. Because of the great technical and theoretical difficulties and uncertainties involved in the genuine cross-cultural testing of abilities, where language, customs, values, and the whole way of life differ markedly between the groups being compared, most investigators in differential psychology have chosen to study different racial groups which share a more or less common culture in terms of language, exposure to formal education, the forms of employment, and the cognitive demands associated therewith. The major racial groups in the United States, at least in recent decades, probably come closest to these criteria.

Numerically, Negroes are the largest of such racial minority groups in the US population. In recent years a good part of the motivation for the psychological study of Negro–white differences in mental abilities has stemmed from the conspicuous and seemingly intractable differences in scholastic performance under fairly equal instructional conditions, and from the relatively large percentage (more than three times that of whites) of Negro youths who fall below the minimum mental qualifications for induction into the armed forces, even when equated with the average white youth in amount of schooling.

Let us review briefly some of the main findings of psychometric research in the two groups that have been compared most extensively, viz. American Negroes and whites.

### Magnitude of the difference

Since mental abilities are seldom measured on an absolute scale, it is customary, for most tests, to describe the units of measurement in terms of the standard deviation of test scores in some representative sample of the population under study. Raw scores (i.e. number right) on mental tests called intelligence tests are usually converted to an IQ scale, with a mean of 100 and a standard deviation of 15 in the normative population.

White–Negro mean differences are most often expressed in units of the standard deviation within the normative population or within the White comparison group, which often amounts to about the same.

The magnitude of the white–Negro test differences in all of the studies reported in the literature vary mainly in terms of several factors: age of the subjects, nature of the test, geographic region, and representativeness of the samples.

*Age of subjects*

Tests devised for assessing the development of children under 2 years of age cannot be called intelligence tests, if by intelligence we mean the general factor common to performance on all complex cognitive tasks in the age groups above 3 or 4 years. Tests of whatever kind administered below 2 years of age show little or no correlation with cognitive tests administered in later childhood and beyond. The infant tests, such as the Gesell, Griffiths and Bayley scales, are reliable measures of early neuro-muscular and sensory maturation and coordination. In the functions measured by those tests, Negro infants are considerably advanced as compared to white infants, up to 15 to 18 months of age. This infant precocity in motoric development has been noted also in a number of studies of African infants, as well as in Negroes in the US. In terms of a developmental quotient, with a mean of 100 and standard deviation of 15, such as provided by the Bayley scale, the white–Negro difference during these early months is of the order of ten to thirty points. The largest differences on record favour African infants and US Negroes in poverty areas in the South. This Negro precocity is also evident in physical indices of skeletal and neurological maturity at birth. There is also some evidence of Negro precocity in the earliest elements of language development, which is intimately related to motoric maturity (documentation of the research on all these points is given in Jensen 1973a).

By 2 years of age, the white–Negro developmental gap disappears. As the mental test content becomes more highly loaded with *g* (i.e. the general intelligence factor which accounts for most of the variance in complex cognitive tests in later childhood and maturity) with each succeeding year, the growth curve of the average white child overtakes that of the average Negro of the same age, and, by 4 to 5 years of age, the difference between the groups, provided the tests are highly $g$-loaded, amounts to about one standard deviation, equivalent to fifteen points on the IQ scale, in favour of the white group. In $g$-loaded tests the white–Negro difference, expressed in standard deviation or $\sigma$ units, does not change after 4 or 5 years of age. I would speculate that this same difference of about $1\sigma$ would be found as far down the age scale as the $g$ factor can be measured. This hypothesis could be tested by comparing the groups in terms of factor scores on the $g$ factor rather than in terms of factorially complex test scores which have a diminishing $g$ component as one moves down the age scale.

The fact that the Negro IQ deficit does not change at all beyond age 5, relative to variation within either the white or Negro group, is of considerable theoretical importance. One of the main pillars of environmentalist explanations of the Negro IQ deficit is expressed by the so-called 'cumulative deficit' hypothesis, which holds that environmental disadvantages act like compound interest in producing a cumulative deficit in Negroes' intellectual development. It has already been mentioned that Negro IQ declines from age 2, when it can first be measured, to age 4 or 5, after which it remains constant. This decline could be due to a cumulative deficit associated with certain environmental lacks, or it could be due to the increasing $g$ loading of intelligence test items between 2 and 5 years of age (by age 5 the $g$ loading of intelligence tests like the Stanford-Binet already closely approaches its asymptotic value). If the deficit were environmental, however, one must wonder why it does not continue to cumulate beyond age 5, when children enter school and are just becoming aware of the social milieu which environmentalists claim contain many of the key ingredients that depress Negro IQ and scholastic performance.

As important as the cumulative deficit hypothesis has been to the environmentalist programme, I have not found any evidence to support it, and much evidence that contradicts it. Most studies of cumulative deficit have failed to control for possible demographic artifacts, such as differences in the populations sampled at various ages. But what is methodologically perhaps the most rigorous study of the subject based on the IQ differences between younger and older siblings within the same families, using all the families in a California school district with children between ages 5 and 12, and controlling for family size and birth order, there was found statistically significant evidence of a progressive deficit in verbal IQ requiring reading ability, but no evidence whatsoever of a cumulative deficit (as indicated by a zero difference between IQs of younger and older sibs) in a non-verbal, highly *g* loaded IQ test (Jensen 1974a). Interestingly, the average white–Negro difference was at least as great on the non-verbal as on the verbal IQ test. The fact that the one standard deviation Negro deficit in non-verbal IQ is stable after age 5 means that its causes, whatever they might be, must be sought in factors whose influences are already fully established before school age.

*Nature of the tests*
The size of the white–Negro difference also depends upon certain properties of the test. Contrary to popular belief, verbal tests do not yield larger differences than non-verbal, and more often the reverse is true. However, my study of this matter leads me to believe that what little difference there is between Negro deficit in verbal and non-verbal tests is not in itself of fundamental significance. Verbal and non-verbal test batteries often reflect varying admixtures of two, more fundamental, classes of abilities, in one of which Negroes show little, if any, deficit, compared to whites, and in the other of which Negroes show their greatest deficit (with the exception of one special ability, viz. spatial visualization). I call these two classes of ability Level I and Level II. Level I consists of abilities such as short-term retention of visual and auditory inputs, memory span, rote learning, and the like. It is characterized by reception, retention, and recall on cue, with a minimum of mental manipulation or transformation. Tests incorporating these features more or less exclusively can be made as demanding and difficult as one likes. They can require every bit as much of the subject's attention and effort as any other kind of test. We have used a variety of such Level I tests in white and Negro samples and find little or no racial group difference relative to the individual variation within groups, which is considerable. Thus, an intelligence test that contains some items which can be acquired merely through familiarity, by repetition or rote learning, such as simple factual information and concrete vocabulary items, will to that extent reflect Level I ability. The Stanford-Binet

and the Wechsler tests include some almost pure Level I tests, such as digit span memory. And to the extent a test is loaded with Level I, it minimizes the white–Negro difference.

Level II ability involves mental manipulation and transformation of inputs in order to arrive at a satisfactory output. This means discrimination, generalization, comparison, planned or goal-oriented search of immediately present stimuli or of stored memories, abstraction, classification, judgement, induction and deduction involving concepts. Level II is much the same as what Spearman termed *g*. The moment any mental manipulation, transformation, selection, or comparative judgement is aroused or demanded by the stimulus input, Level II or *g* enters the picture as a source of individual differences in the response. It is, of course, a greater source of variance the more the task calls for Level II processes relative to other sources of variance, such as Level I processes, sensorimotor abilities, attention, effort, and the like.

Test items that call for problem solving with novel materials, as contrasted with items that require recognition or recall of previously learned material, are the best measures of Level II, and they are the items with the highest *g* loadings when tests are subjected to factor analysis. Items such as those found in Raven's Progressive Matrices test are almost pure Level II, for example, while digit memory (i.e. repeating a string of digits immediately after hearing them spoken at a 1 second rate) is almost pure Level I. As soon as we introduce some mental manipulation into the memory task, however, it takes on some Level II loading. It has been found, for example, that in a factor analysis of a number of Level I and Level II tests, forward digit span had nearly all of its factor loading on the Level I factor, while backward digit span (i.e. reciting the digit series in reverse of the order of presentation), had its factor loadings divided between the Level I and Level II factors, with slightly more on the latter. White and Negro groups differ most on the Level II factor and little, if at all, on Level I (Jensen 1970 1971a 1973c 1974b).

A thorough survey of 382 studies involving some eighty different standardized intelligence tests in whites and Negroes shows an average difference of about one standard deviation; the great majority of the group mean differences are between ten and twenty IQ points (Shuey 1966). All of these tests are predominantly *g* loaded, but many include other factors as well.

Attempts to show difference in the ability profiles of whites and

Negroes on tests of verbal, numerical and figural reasoning and the like (e.g. Lesser, Fifer and Clarke 1965), I strongly suspect, are merely derivative, secondary phenomena reflecting the different Level I and Level II demands of the various tests. The available evidence does not appear to me to support the interpretation that whites and Negroes have different profiles in the so-called primary mental abilities themselves, except in so far as measures of these abilities cannot be divorced from their Level I and Level II demands. But there is one important exception, i.e. spatial visualization ability.

A number of studies suggest that Negroes perform further below other groups (whites, orientals, American Indians, Eskimos) on tests of spatial visualization ability than on tests of any other ability. This has been found in Negroes of the West Indies as well as of the United States. The same tests given to African Negroes show even lower scores, but they are not appreciably lower than a variety of $g$ loaded tests which do not require spatial ability.

Spatial ability has long been suspected of being sex-linked, since it is the only one of Thurstone's seven primary mental abilities which consistently shows an appreciable sex difference. Only about one-fourth of females exceed the male median in tests of spatial ability. Since Bock and Kolokowski (1973) have now demonstrated by quantitative genetic analysis that spatial ability is influenced by a single X-linked recessive gene, it is important from the genetic standpoint to see if this fact can help to explain the findings on spatial ability in American, West Indian and African Negroes, and on the direction and relative magnitudes of sex differences in spatial ability, as compared with other abilities, in Negro and white groups. The present evidence, such as it is, appears consistent with the X-linkage of spatial ability and the additional fact that the 20 to 30 per cent admixture of Caucasian genes in American Negroes came largely from male white ancestors, thereby resulting in the introduction of proportionally about one-third fewer X-linked than autosomal Caucasian genes into the American Negro gene pool. A rigorous test of this genetic hypothesis, however, awaits additional data (Jensen 1975). But it is of interest that quantitative genetics already has a theoretical model, in the mechanisms of X-linkage and recessivity, that appears capable of predicting the findings on white–US Negro differences in spatial ability and their interaction with sex differences in spatial ability. Environmentalist explanations of these facts at present would have to be especially *ad hoc*, and would probably encounter difficulty with the fact that spatial ability, unlike *g* loaded tests, has relatively little correlation with socio-economic status within racial groups.

Tests of scholastic achievement generally show slightly smaller white–Negro differences than most standard intelligence tests. This seems surprising to many but is consistent with the idea that some scholastic knowledge and skills, such as spelling and mechanical arithmetic, are partly acquired by Level I processes. Scholastic tests which require the student to reason with his specific knowledge and skills to solve novel problems, however, are very highly correlated with general intelligence tests, and even with non-verbal tests of *g*, when all the testees have had the same number of years of schooling.

*Geographical region*
The nationwide testing of youths for induction into the armed forces clearly reveals regional differences in intellectual ability, both for whites and Negroes, though the regional differences are considerably larger for Negroes than for whites. The white–Negro differences in various regions vary from the overall White average the equivalent of about ten to twenty IQ points. Negro IQs are lowest in the South and Southeast and there is a gradient of increasing IQ as one moves further north and west. There is a similar, though less pronounced, gradient of IQ in the white population. This regional variation in IQ appears to be mostly a result of past selective migration associated with economic factors and employment opportunities making different educational and intellectual demands. It is of interest from our standpoint that variation in the amount of Caucasian admixture in American Negroes follows much the same regional gradient as IQ variation, from the Deep South, with close to 10 per cent Caucasian admixture, to the North and West, with about 20 to 30 per cent, and the Northwest as high as 40 per cent (Reed 1969). Since practically all the Caucasian genes in the American Negro gene pool were introduced during the period of slavery, which was confined to the South, the present regional variation is undoubtedly due to selective migration. It is significant that IQ and amount of Caucasian admixture in Negroes parallel one another in geographical distribution, and that both of these variables more

or less parallel the regional variations in the IQ in the white population.

*Representativeness of the sample*

White–Negro comparisons have been reported where one or both groups are atypical samples of the white or Negro populations of a particular locality. Comparisons of white and Negro prisoners, juvenile delinquents, and patients in public hospitals, are examples. Such biased samples usually reduce the racial difference. The most frequent type of biased sampling is the matching of the racial groups on some index of socioeconomic status (SES), such as income and occupational and educational level. Such matching of the racial groups generally reduces their IQ difference by about one-third of a standard deviation, more or less, depending on how many IQ-correlated factors enter into the matching. It also depends, in the case of children, on whether one matches Negro and white children at the upper or at the lower end of the SES scale. High SES Negro and white children differ more in IQ than groups matched for low SES. In a review of the thirty-three studies before 1965, including a total of about 7900 Negro and 9300 white subjects, in which white and Negro groups were of comparable SES, Shuey (1966, p. 520) concluded:

> The consistent and surprisingly large difference of 20·3 IQ points separating the high-status whites and high-status coloured is accentuated by the finding that the mean of the latter groups is 2·6 below that of the low-status whites. It is probable that the home, neighbourhood and school environments of the white and coloured lower-class children tested are more nearly alike in their stimulating qualities than are the home, neighbourhood, and school environments of the white and coloured upper and middle class children; but it seems improbable that upper and middle class coloured children would have no more cultural opportunities provided them than white children of the lower and lowest class.

Three more recent studies involving large samples also found low SES white children to have slightly higher IQs than middle and upper SES Negro children (Wilson 1967, Scarr-Salapatek 1971, Jensen 1974b).

## The hypothesis of culture-biased tests

The most popular explanation of these psychometric differences, in whole or in part, is that the tests are in some way biased so as to favour whites and disfavour Negroes. Since the tests often have been standardized on the white population, it is claimed that they are culturally loaded with content peculiar to Anglo middle class experience, although this has certainly not been the intention of test constructors.

The claims of cultural bias as an explanation of the white–Negro IQ difference in the United States runs into numerous difficulties. For one thing, many of the tests that show the greatest white–Negro difference show much smaller differences for other minority groups which are also regarded as disadvantaged or culturally different. On non-verbal IQ tests, which do not handicap children brought up in a foreign tongue, American Indians and Mexican-Americans outperform Negroes, on the average. On Raven's Progressive Matrices, one of the best highly $g$-loaded non-verbal intelligence tests, Arctic Eskimos, with their extremely different culture, score at least up to the white norms obtained in Scotland and the US Chinese and Japanese in the United States at present score at least as high as native whites, and in California they score higher, especially on highly $g$ loaded non-verbal tests. Moreover, no one has yet devised or standardized an intelligence test within the Negro population which significantly narrows the racial IQ difference, although there have been serious attempts to do so. Yet most intelligence tests originating in the United States can be used in foreign countries simply by translating the test instructions and verbal items into the appropriate language. The translated tests retain highly similar reliability, validity, inter-item correlations and score distributions as are obtained in the US white population. This has been the usual experience with the Stanford-Binet and Wechsler tests, which have been used in many countries with seldom more than translation and substitutions of a few of the informational items, such as changing 'What is the population of the United States?' to 'What is the population of Japan?' in the Wechsler test. A translation of the Stanford-Binet test into Negro ghetto dialect, however, produced no significant increment (one IQ point, in fact) over the IQ obtained with the standard English version when given to Negro children most familiar with the ghetto dialect.

Recently, I have conducted intensive studies of culture bias in tests, using large samples of typical white, Negro and Mexican children in California schools. I will here summarize the main results.

But first of all, one must distinguish between culture *loading* and culture *bias*. A test may contain informational content that could only be acquired within a particular culture. This can usually be determined simply by examination of the contents of the test items. Whether the particular cultural content causes the test to be biased with respect to the obtained scores between any two groups is a separate question. If the test includes only cultural content that is common to the experience of the groups being compared, it will not be culturally biased, assuming that the testing procedure itself is not a source of bias.

The fact that racial and social class groups differ on a test cannot itself be a proper criterion of bias. Legitimate criteria of test bias are of two types: external and internal. External bias is related to the predictive validity of the test, i.e. how well it predicts such criteria as school grades, success in some specialized training, and occupational performance. A test is biased if the intercepts and slope of the regression of criterion measures on test scores differ significantly for the two or more populations in question. Reviews of the research on this point comparing white and Negro samples are unequivocal with respect to scholastic and job performance. There is negligible difference in the slopes and intercepts of regression lines for white and Negroes. A single regression equation predicts equally well for both groups (Humphreys 1973, Linn 1973). Interestingly, the few exceptions reported in the literature would favour the Negro groups if the tests were used for selection, i.e. the difference in the regression lines is such that for any given test score whites slightly out-perform Negroes on the criterion. In brief, the overwhelming evidence on the predictive validity of standard tests indicates that they are not biased against Negroes when compared with whites (there are too few studies of other ethnic groups to permit any general conclusions about them).

It can, of course, be argued that the criterion predicted by the test scores is itself culture biased, and that one therefore needs a culture biased test to predict a culture biased criterion (e.g. scholastic achievement). Therefore, one must consider various internal criteria of test bias. These internal criteria seem especially appropriate for investigating the hypothesis that a given test is biased for one population when the item selection and standardization were based on a different population. If the test items are culture loaded, i.e. they call for specific information acquired in a particular culture, and if the cultures of the standardizations and target groups differ with respect to the cultural information sampled by the items, this should be reflected in the various internal indices of bias. I will list each of these indices and describe what we have found concerning each one with respect to white–Negro comparisons (Mexican-American children were included also, but for the sake of simplicity I will not attempt here to summarize these results; in general, they differ little from the results for Negroes, except that the Mexican subjects do relatively better on the non-verbal matrices test and relatively worse on the picture-vocabulary test). All of these analyses have been made on what is probably the most culture loaded of all standard intelligence tests, the Peabody Picture Vocabulary Test (PPVT), and on one of the least culture loaded tests, Raven's Progressive Matrices.

The PPVT consists of 150 plates, each with four pictures. The examiner names one of the pictures and the subject is asked to point to it. The vocabulary ranges from very easy, common, and concrete words to very rare words and abstract concepts. The Progressive Matrices consists of sixty plates, each with a missing part which the subject must select from a multiple-choice set of six to correctly complete the pattern. Items range in complexity and difficulty from a level that is passable by most 3-year-olds up to a level of difficulty beyond the capacity of the average adult. Figure 13.2 shows typical PPVT and Raven items of moderate difficulty.

The subjects in these studies numbered more than 3000 children in California schools, about equally divided among the racial groups (I have presented these studies in detail elsewhere: Jensen 1974c).

(1) Correlation of raw scores with chronological age in months does not differ appreciably for whites and Negroes on either the PPVT or Matrices.

(2) Internal consistency reliability and average inter-item correlation of both tests are the same for Negroes and whites.

(3) Rank order of item difficulty (as indicated by per cent passing) is virtually the same in both racial groups. The correlation between $P$ values (i.e. per cent passing an item) over

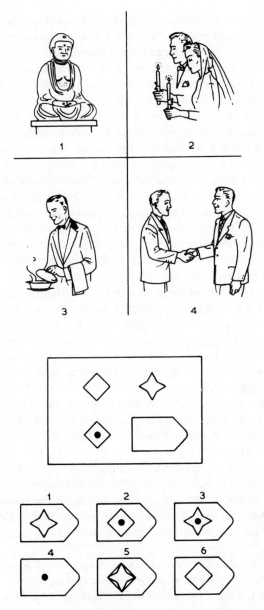

*Figure 13.2* Sample items of the Peabody Picture Vocabulary Test (top) and Raven's Progressive Matrices. The PPVT word for this item is 'ceremony'.

all items for whites and Negroes is near perfect, without correction for attenuation (for PPVT $r = 0.986$; for Matrices $r = 0.993$). When the correlations are obtained for various subsets of twelve or fifteen items, they are still very high (0.87 to 0.99), and the correlations are highest in those subsets of items that discriminate most (i.e. have the largest differences in $P$ values) between the races. This is the opposite to what one should predict from a culture bias hypothesis of the group differences, which should lead to the expectation that the most discriminating items would show the least similarity between the groups in the rank order of $P$ values. In many subsets of items the correlation of $P$ values between races is higher than between boys and girls within the same race, although boys and girls score about equally, overall. Certain PPVT items show more sex bias than any items show racial bias. For example, 'parachute' versus 'casserole' reflect different sexual biases in cultural knowledge. The PPVT also reveals culture biases in comparing white school children in England and white children in the United States. Although both groups obtain about the same total score, some vocabulary items are much easier for the English than for the Americans (e.g. 'pedestrian' and 'goblet') and vice versa (e.g. 'bronco' and 'thermos'). Negro and white groups in California schools, on the other hand, do not show any of these marked discrepancies in order of item difficulty.

(4) An even more sensitive index of cultural differences is the correlation between the item '$P$ decrements' for the two races. The $P$ decrement is the difference between the per cent passing two adjacent items, e.g. $P_1 - P_2$, $P_2 - P_3$, etc. Thus we are measuring the racial group similarity in the differences in difficulty among items. Again, these correlations are very high (0.79 for PPVT, 0.98 for Matrices), and again the correlation was highest for the most discriminating sets of items. Correlations between the sexes within racial groups are not significantly greater.

(5) Items that best discriminate individual differences *within* racial groups (i.e. items with the highest correlation with total test score) are the same items that discriminate most *between* the racial groups.

(6) Incorrect response (errors) are distributed in a non-chance fashion over the multiple-choice distractors in the same proportions for whites and Negroes. There were several significant

exceptions to this in the Matrices; that is, on some items Negroes made different errors than whites. However, in every such instance it was found that the Negro children's proportions of responses to the various error distractors were the same as the proportions for white children who were approximately two years younger in chronological age. Thus it appears that the few differences that were found between white and Negro children are most clearly related to differences in level of mental maturity than to cultural differences.

(7) The matrix of inter-item correlations for each test was factor analysed within each racial group to determine the loadings of each item on the general factor (i.e. first principal component) that accounts for most of the covariance among all the items. The items' factor loading for Negroes and whites are highly correlated, and, most significantly, the correlation is markedly increased when the Negro factor loadings are correlated with the factor loadings of whites who are about two years younger. In fact, on this index, 6th grade (ages 11–12) Negroes are more like 4th grade (ages 9–10) whites than like 4th or 5th grade Negroes (also 6th grade Negroes obtained about the same total raw score as 4th grade whites). Moreover, the loadings of items on the general factor *within* each racial group show a high positive correlation with the degree to which the items discriminate *between* the races. In other words, those items which best measure what is common to all items *within* each race are the same items that show the largest race difference.

(8) Few if any psychologists would claim that Raven's Matrices is more culture loaded than the PPVT. If the PPVT is culturally biased against Negroes, then, if we perfectly match PPVT and Matrices items for difficulty (i.e. per cent passing) in the white population, we should expect, from the culture bias hypothesis, that these two sets of items would not be matched in difficulty in the Negro population. For Negroes, the culturally loaded PPVT items should be more difficult than the Matrices items. But, in fact, we found no significant difference. Thirty-five PPVT and Matrices items which are perfectly matched in difficulty for whites turned out to be matched in difficulty for Negroes as well (this was not true of Mexicans, for whom the PPVT items are significantly more difficult, as would be expected from the culture bias hypothesis).

(9) Finally, using an analysis of variance to examine the Race

× Items interactions (for both PPVT and Matrices), we found we could almost perfectly *simulate*, without statistically significant differences, all features of the Negro–white differences, using entirely white samples. We simply divided the entire white sample into two groups, a younger group (ages 6 to 9) and a slightly overlapping older group (ages 8 to 11). Detailed comparisons of these two groups simulate, within the margin of sampling error, the results of the same comparisons of whites and Negroes, when both groups are of the same chronological age. We have found no feature of the PPVT or of the Matrices which distinguishes Negroes from whites who are about two years younger, or which distinguishes any differently between Negroes and whites of the same age than between groups of younger and older whites.

All these findings seem to me very incompatible with the culture bias hypothesis. To maintain this hypothesis one would have to postulate the additional and supremely *ad hoc* hypothesis that the cultural differences between Negroes and whites perfectly simulate age differences within the white group, with respect to item difficulties, $P$ decrements, inter-item correlations, choice of distractors, and $g$ factor loadings, for tests as diverse as the PPVT and the Matrices.

In another study in which several mental tests were administered to several thousand white and Negro children by twelve white and eight Negro examiners, it was shown that the race of the examiner had no significant or systematic effect on the intelligence test scores of white and Negro pupils (Jensen 1974d). Also, special tests devised to measure attention, speed, persistence and effort in the testing situation revealed only negligible differences between Negroes and whites. I therefore conclude that these factors are an unlikely explanation of the large race difference in intelligence test scores.

Another study has shown that administering several mental tests under speeded conditions versus no time pressure did not significantly alter the white–Negro difference, although both groups performed better under the more lenient condition (Dubin, Osburn and Winick 1969). The same study also showed that pretest practice on alternate forms of the tests did not significantly reduce the racial differences.

The most reasonable hypothesis, it seems to me, is that the two racial groups differ in the rate and the asymptote of development of the brain processes underlying the general factor common to intelligence test items. Comparisons of the racial groups across the ages from early childhood to adolescence on a number of different indices of mental growth lends further support to this hypothesis.

## Consistency among development indices
The Gesell Figure Copying Test (Ilg and Ames 1964, pp. 63–129) consists of the ten geometric forms shown in Figure 13.3. The subject is encouraged simply to copy each figure, without time

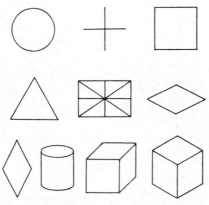

*Figure 13.3* The ten simple geometric forms used in the Figure Copying Test. In the actual test booklet each figure is presented singly in the top half of a 5½ x 8½ inch sheet. The circle is 1¾ inches in diameter.

limit. A pencil with an eraser is an essential part of the testing procedure. The test approximates a Guttman scale, i.e. it is like a series of hurdles, in that, if a subject can correctly copy, say, the fifth figure in the series, in all probability he can copy correctly all the preceding figures; and if he cannot correctly copy, say, the sixth figure, he will in all probability fail all the figures that follow it. The test reflects mental development over a range from about age 3 to age 12. When the Figure Copying Test has been factor analysed along with standard intelligence tests, it is loaded almost entirely on the *g* factor.

We have given this test to more than 10,000 school children of different ethnic groups, of ages 5 to 12 years. There are marked group differences at every age, with orientals scoring highest, followed closely by whites, then Mexicans and, lastly, Negroes. The magnitude of the difference on this test is almost two standard deviations between orientals and Negroes, as can be seen from Figure 13.4. Negro children in the 4th grade (ages 9–10) perform on a par with oriental children in the 1st grade and slightly below white children in the 2nd grade. The Mexican group, although lowest in socioeconomic status, is almost exactly intermediate between the orientals and Negroes and nearly on a par with whites.

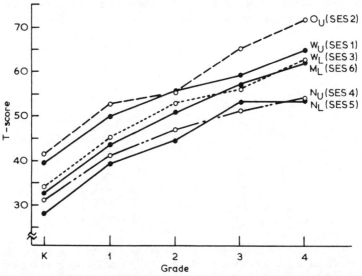

*Figure 13.4* Oriental (O), white (W), Mexican (M) and Negro (N) groups from socio-economically urban, largely middle to upper-middle class (U) and rural, largely lower to middle class (L) communities. The six groups are ranked from highest (SES 1) to lowest (SES 6) on a composite index of socio-economic status.

Even more telling is the fact that all these groups show the same developmental sequence of difficulties in copying these figures. The same conceptual difficulties appear in all the various ethnic groups, but simply at different ages, on the average. The difficulties of Negro children of ages 6 or 7 are indistinguishable from the difficulties of white and oriental children of ages 5 or 6. Each figure, so to speak, 'evolves', going from younger to older ages. Typical examples of some of the modal difficulties, as one goes from drawings of lesser to greater maturity, are shown in Figure 13.5. It would seem hard to explain in terms of cultural differences why Negroes, whites, orientals, and Mexicans all go through the same sequence of these peculiar characteristics of copying figures, and differ only in the average age at which they encounter the various difficulties up till the age at which they are able to copy the given figure correctly.

Jean Piaget has devised a number of highly diverse developmental tasks with similar properties. They are seemingly simple tasks, utilizing familiar objects, which call for judgement, mental manipulation, and reasoning about matters universally available to

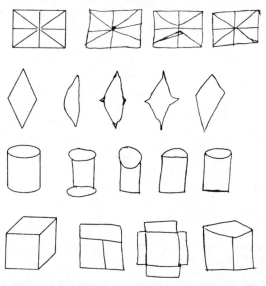

*Figure 13.5* Typical examples of developmental changes in children's figure copying, going from lesser to greater maturity. The models are on the extreme left.

observation. One example is shown in Figure 13.6. The child is shown a bottle of red liquid (A). An opaque card is then placed in front of the bottle, and the bottle is tilted (B). Then the child is given a full-scale outline drawing of the bottle and is asked to draw the level of the red liquid as it will appear when the card is removed (C). Most children under 8 or 9 years of age draw a line more or less parallel to the bottom of the bottle, as shown in C, while older children more often correctly draw a horizontal line. Many children under 8 years do not markedly improve their drawing even after they have been shown the liquid in the tilted bottle. When this water-level test was given to large representative samples of three ethnic groups in Grades 1 to 3 (ages 6–8) in California schools, the per cent of each group passing the test was :

oriental, 43 per cent, white 35 per cent, Negro 1.3 per cent. Tuddenham (1970) gave nine other such Piagetian tests of different concepts to the same groups. Negroes did less well than whites on every item; oriental children exceeded white children on seven of the ten items. The differences are comparable to those found with highly $g$ loaded tests such as Raven's Matrices (I have elsewhere reviewed in greater detail these and other studies showing similar results: Jensen 1973a, 312–18).

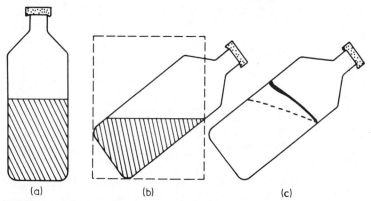

(a)                          (b)                          (c)

*Figure 13.6* Piaget's tilted water bottle test, to measure the concept of the horizontality of water level.

Another developmental index is interesting because it has no right or wrong answers, but only preferences which change systematically with age. As children mature mentally, they show changing preferences for colour, form, number and size, in that sequence, in attending to the attributes of objects. The order of preference for children of kindergarten age (5 or 6 years) is (1) form, (2) colour, (3) number, and (4) size. Groups of white and Negro kindergarten children were each shown twelve different stimulus displays of the type shown in Figure 13.7.

The figure on the four cards differs simultaneously in colour (green, red, blue, yellow), shape, size and number. The examiner gives the small card at the top to the child and aks him to put it down on any one of the four cards with which he thinks it goes best. It is made clear that there is no 'right' answer. Thus, the child can match the target card on the basis of colour, form, number or size. It turned out that white and Negro children of the same age differed in the relative frequencies of their preferences,

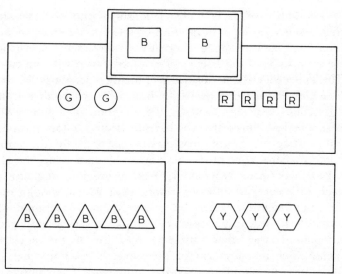

*Figure 13.7* An item of the colour-form-number-size test. The letters (which do not appear on the actual test) indicate the colours of the figures.

in accord with the developmental prediction, i.e. the Negro children had a significantly higher percentage of the less mature preferences (Toki 1971):

|  | Preference per cent | | | |
| --- | --- | --- | --- | --- |
|  | Colour | Form | Number | Size |
| White | 10·2 | 73·5 | 14·9 | 1·5 |
| Negro | 28·9 | 56·5 | 12·5 | 2·1 |

In view of the consistent racial differences in these highly diverse developmental tasks. I venture the hypothesis that typical Negro and white children will show consistent differences in growth rates and in orderly sequential development on all age-related tasks involving abstraction, judgement, mental manipulation – in short, the essence of $g$ or Level II ability.

## Information processing and the essential nature of $g$

Examination of the most highly $g$ loaded test items shows them to be clearly characterized by their requiring the testee to process a considerable amount of information – information not in the sense of merely recalling stored knowledge, but in the sense of having to take a number of facts and relationships into consideration simultaneously in order to produce or select the correct answer. Complexity, choice among alternatives, judgement, decision – these seem to be of the essence of *g*, as contrasted with memory, factual knowledge and performance of highly practised skills.

The idea of *g* as being related to information processing in this fundamental sense has come to the attention of a number of experimental psychologists, who have devised laboratory measures of information processing capacity. In order precisely to quantify the informational content of a task on an absolute scale, the task has had to be made very simple. And in order to demonstrate reliable differences in difficulty among tasks which, though they differ only slightly, differ by precisely known amounts in informational load, it is necessary to employ a very sensitive and continuous (rather than discrete) measure of the subject's response. The measurement of the subject's reaction time (in milliseconds) to stimuli meets this requirement. The complexity of the stimulus situation is varied so as to convey different amounts of information as measured in *bits*, the unit of measurement in information theory. A bit (for *binary digit*) is the amount of information that reduces uncertainty by $\frac{1}{2}$. A stimulus situation to which the response is completely predetermined and therefore requires no discrimination, comparison, judgement or decision has no uncertainty and therefore conveys zero *bits* of information. The next most complex stimulus situation, involving two elements or alternatives, requires one decision and has one *bit* of information, e.g. Yes or No. Four alternatives have two *bits* of information, e.g. One or Two? Yes; One? Yes. Eight alternatives have three *bits* of information, and so on. The number of *bits* of information is the logarithm, to the base 2, of the number of alternatives. Very complex, highly *g* loaded test items undoubtedly contain many *bits* of information, but the actual number for any given item is not determinable by any means presently known. But using vastly simpler stimuli, though they are not nearly as good a measure of *g* because of their much smaller informational content, permits exact quantification of the task's complexity in terms of *bits*.

One of the important discoveries in this field, often called Hick's Law (Hick 1952), which has been replicated in many studies, is that the subject's reaction time (RT) increases as a linear function of the amount of information as measured in *bits*.

Thus, choice RT (i.e. responding differentially to two or more stimulus alternatives) is invariably greater than simple RT (i.e. response to a single stimulus). Hick's Law has been demonstrated by a variety of laboratory techniques, using different stimuli and different sensory modalities.

There are highly reliable individual differences in simple and in choice RT. Though it is not of much practical importance, it is of great theoretical significance that individual differences in simple RT are not significantly correlated with scores on standard intelligence tests, while choice RT is correlated with intelligence (Eysenck 1967). The correlation is negative, i.e. the more intelligent subjects take less time to process a given amount of information.

We have devised an apparatus for very precise measurement of RT, as well as movement time (MT), in response to stimulus arrays varying in informational content from zero to three *bits*. This is an extremely small range of difficulty, so small, in fact, that persons have little or no subjective feeling that the three *bits* task is any more difficult than the zero *bit* task. Even the experimenter cannot perceive a difference in the subject's RT in the zero *bit* and three *bits* tests. But when precisely measured by electronic timers, RT increases regularly by some 30 to 50 milliseconds with each additional *bit* of information conveyed by the stimuli. These small fractions of a second, however, are subjectively negligible to subjects. This is very unlike ordinary intelligence tests in which the items often increase very perceptibly in complexity and difficulty, even to the point that the increasing appearance of difficulty can possibly intimidate subjects and discourage continuing effort.

The apparatus for measuring the subject's RT and MT consists of a panel, 13 × 17 inches, painted flat black, and tilted at a 30° angle. At the lower centre of the panel is a red pushbutton, $\frac{1}{2}$ inch in diameter, called the 'home' button. Arranged in a semicircle above the 'home' button are eight red pushbuttons, all equidistant (6 inches) from the 'home' button. Half an inch above each button (except the 'home' button) is a 1 inch faceted green light. Different flat black panels can be fastened over the whole array so as to expose arrays having either 1, 2, 4, 6 or 8 light-button combinations.

The subject is instructed to place the index finger (of his preferred hand) on the 'home' button. Then an auditory 'ready' signal is sounded (a high-pitched tone of 1 second duration), followed, after a continuous random interval of from 1 to 3 seconds, by one of the green lights going 'on', which the subject must turn off as quickly as possible by touching the sensitive microswitch button directly under it. RT is the time the subject takes to remove his finger from the 'home' button after the green light goes on. MT (movement times) is the interval between removing the finger from the 'home' button and touching the button which turns off the green light. RT and MT on each trial are registered in milliseconds by two electronic timers. On each trial never more than one light in the whole array goes 'on', and the subject turns it off by touching the button adjacent to the light. The particular light that goes 'on' in each trial is completely random and thus unpredictable by the subject, thereby creating the uncertainty upon which the quantification of information depends.

Our experiments with this apparatus have shown, in accord with Hick's Law, that RT increases as a perfectly linear function of *bits* of information, in school children and in young adults. The linear increase in RT as a function of *bits* shows up for individuals as well as for the group as a whole, and is therefore a very lawful and reliable psychological phenomenon. The average correlation between RT and *bits* for *individual subjects* is over 0·9, which means that even for individuals (and not just for the group average) there is an almost perfect linear regression of RT on *bits* of information. MT, on the other hand, is completely unrelated to *bits* and remains constant across all amounts of information, both for individuals and for the group means. The reliabilities of both RT and MT are about 0·90 when subjects are given thirty trials on each light–button combination. Also, it is apparent that RT and MT are not measuring the same sources of individual differences variance, since the correlation between the two measures is only about 0·3. Moreover, there is virtually no functional relationship between RT and MT, as indicated by a *within*-subjects correlation between RT and MT of close to zero.

A measure of information processing capacity that is independent of absolute RT is the slope, *b*, of the linear regression of RT on *bits*. When the regression of RT on *bits* is determined for every subject, the values correlate significantly but lowly (around −0·30) with standard intelligence test scores. This accords with the hypothesis, as described by Eysenck (1967), that information processing capacity, assessed independently of absolute RT in terms of the rate of increase in RT as a function of the increasing complexity of the task, is a significant component of intelligence.

Now what has all this to do with our topic of race differences? In a recent study we hypothesized the following: if the essence of the $g$ component in the subject's performance is related to the degree of complexity of the task (in the information theory sense), and if whites and Negroes differ in $g$ capacity, then whites and Negroes should show no significant difference in performance on tasks of zero information (e.g. one light–button combination or simple RT), but should show increasing differences as the number of *bits* of information increases, even when the increases in information are all within such a narrow range and at such a low level of complexity (i.e. between zero and three *bits*) as to be subjectively indiscriminable in difficulty.

We tested this hypothesis on 200 male youths, 18 to 19 years of age, with nearly equal numbers of Negroes and whites. The samples were not representative of the general population. All subjects were within the normal range of intelligence and the white and Negro groups were almost perfectly matched in the score distributions on a group verbal test of general intelligence. They were also matched as closely as possible in years of schooling (averaging 11·5 years), although the Negroes averaged about half a year *more* schooling than the whites. Many studies have matched racial groups on socioeconomic, educational and other environmental factors correlated with intelligence test scores, and have shown that the group difference is diminished by such matching, often with the claim that if more such environmental factors had been controlled, the test difference would be wiped out completely. The present racial samples go a step further: they do not differ appreciably in test scores. This stacks the cards against our hypothesis that the groups would differ increasingly in an information processing measure as the amount of information increases. Negro and white groups which score the same on more or less culturally and educationally loaded verbal paper-and-pencil tests, if measured on a series of much less culturally and educationally loaded tests involving regularly increasing amounts of information corresponding to $g$, should show regularly increasing differences as the amount or information increases. And this should be shown to occur even within such a narrow and easy range of cognitive difficulty as not to be subjectively perceptible by the subjects, so that attitudinal and motivational factors would be a most unlikely cause of any differences. Also there is no *a priori* reason to believe that such factors would affect RT for different amounts of information independently of MT (movement time). We already knew from previous studies that information processing was related to RT but not to MT, which is uncorrelated with the informational content of the task.

The results turn out perfectly in accord with the hypothesis. The white and Negro groups differed negligibly and non-significantly (about 3 milliseconds) in mean and median RT to zero *bits* (i.e. the one light–button task), but the white–Negro differences in RT increased significantly and linearly with each additional *bit*, at the rate of 10 milliseconds per *bit*. At three *bits* the groups differ 31 milliseconds in RT, a highly significant difference ($P < 0.001$). The slope of the regression of RT on *bits* was determined for each individual, and the difference between the mean slopes of the white and Negro groups was shown by a *t* test to be significant ($P < 0.01$), with the Negro group showing the steeper slope, i.e. greater increments in RT as the *bits* of information increase (note that the slope measure is independent of absolute RT). The increase of RT as a function of *bits* shows no significant departure from linearity in either group. Slope also correlated significantly (about 0.30, $P < 0.01$) with the mental test scores within groups. Also, it is interesting that intra-individual variability (i.e. an individual's variation about his own mean over repeated trials) increases systematically as a function of *bits*. The rate of this increase was very significantly greater for the Negroes; in fact, it was the largest racial difference to show up in any of the measures derived from this testing procedure.

The groups also differed significantly on MT (white < Negro), but MT showed no correlation with *bits*.

The multiple correlation, $R$, between several of the RT and MT measures, on the one hand, and the racial dichotomy, on the other, is 0.41.

The important and indisputable point of this study is that the two racial samples, which differ much less (in fact, not significantly) in ordinary psychometric scores than do the general populations of whites and Negroes, and in which the Negroes have more education than the whites, still show highly significant differences in a behavioural task, in accord with prior expectations based on theoretical considerations of the essential nature of *g* and information processing capacity.

The findings, however, are not without precedent. In unselected samples of Negroes and whites, even larger differences have been found in choice RT. Noble (1969) found a highly significant ($P < 0.001$) difference between Negro and white children (matched for age and sex) on a four-choice RT test. In a sensitive measure of speed of visual information processing (requiring no motor response at all), using a visual recognition test involving only two *bits* of information (i.e. four alternatives), Bosco (1970) also found a highly significant difference between a group composed mostly of low SES Negro children and a group of middle SES white children. Poortinga (1972) measured simple and choice RT to both auditory and visual stimuli in groups of native African and European students in South Africa (the groups differed $2 \cdot 89\sigma$ on Raven's Matrices.) Choice RT (two and three *bits*) for both the visual and auditory stimuli showed the Africans to have significantly longer mean RTs; in units of the white group's standard deviation, the white–African difference was $1 \cdot 9\sigma$ for auditory and $1 \cdot 5\sigma$ for visual RT. But there was *no difference* between the groups in simple RT (i.e. zero *bits*). In terms of our hypothesis, however, some doubt is raised about the interpretation of these striking results by the fact that visual and auditory choice RT showed negative though non-significant correlations with Raven's Matrices in the African sample and a significant correlation ($-0 \cdot 45$) only for auditory choice RT in the white sample. Such puzzles, of course, simply indicate the need for further experimental analysis.

## Environmentalist hypotheses

Innumerable hypotheses have been put forward to explain the white–Negro difference strictly in environmental terms. Those which have been formulated clearly enough to be tested have not stood up when put to the test. I have reviewed the claims of environmentalists about each of numerous factors said to explain the lower IQ of Negroes – inequality of schooling, teacher expectancy, motivation, language deprivation, nutrition, and reproductive casualty. None of them adequately accounts for the facts of Negro performance on mental tests (Jensen 1973a 1973b). As researchers find each of the more obvious environmental factors, such as those associated with SES, not to hold up as explanations of the Negro IQ deficit, more subtle, often unmeasurable, environmental influences have been hypothesized. In the past few years, each newly proposed environmental hypothesis has failed as soon as it was put to the test.

One of the most popular of recent hypotheses has been that the majority of Negro children have a different language than standard American English, and this supposedly handicaps them in scholastic performance and in taking IQ tests. But how would this explain the results on non-verbal test? And why do immigrant children with little or no knowledge of English not show a similar deficit? Why do children who were born deaf and are therefore severely language-deprived and score lower on verbal tests show no deficit on non-verbal tests? A recent comprehensive review of the research evidence pertaining to the 'different language' hypothesis found no support for it – 'In general, no acceptable, replicated research has found that the dialect spoken by black children presents them with unique problems in comprehending standard English' (Hall and Turner 1974, p.79). The investigators believe the explanation of the Negro deficit must be sought elsewhere, stating that they 'are convinced that more effort should be directed toward studying universals of cognitive development rather than toward relatively superficial performance differences such as spoken dialects' (p. 80).

Another popular environmental hypothesis is that the cause of the Negro IQ deficit is to be found in the quality of the mother–child interaction during the preschool years. The hypothesis is admittedly difficult to test, since appropriate investigation must rely upon naturalistic, systematic, comparative observations of Negro and white children in their natural psychological environments. Two developmental psychologists, Alfred and Clara Baldwin (1973), have spent more than a decade conducting this kind of investigation, including several hundred records of mother–child interactions involving preschoolers in both Negro and white families from lower and middle social class. Many aspects of mother–child interaction (thirty-five coded variables) were systematically observed and recorded in half-hour long free-play settings. Only one significant ethnic difference showed up: Negro mothers were *more* likely than white mothers to adopt a didactic teaching role in free play. The Baldwins note that 'White mothers were much more relaxed in general about the child's academic future. They felt considerably less pressure to teach him academic-type facts during the play session than did the black mothers' (Baldwin and Baldwin 1973, p. 720). They continue:

> On no other measures did we find ethnic differences. The amount of interaction was not consistently different for the black and white groups; the level of syntactic complexity was not different if educational level is held constant. Except for the fact that didactic teaching does involve more direct behaviour requests, we saw no evidence that black mothers were more bossy or more punitive. In fact, we observed very little punitiveness in any of the play sessions.

In the light of their observations, the Baldwins believe the language deprivation theory is called into question: 'All these facts lead us to question deeply whether there is any social significance in the small difference in the syntactic complexity found in the mothers in the free-play session'. They admit: 'Frankly, when we began this investigation, we anticipated many more differences between the black lower class sample and the white upper middle class sample. . . . But as we observed these mother-child pairs, and then as we saw the results of the data analysis, we have become convinced that the most striking fact is the overall similarity of mother–child interaction in free play in all the samples' (p. 720).

In view of the failure of numerous environmental hypotheses to be borne out by evidence, the genetic hypothesis appears reasonable and highly likely, which is not to say that it is proven. But at least it is already established that genetic factors are the most important determinant of IQ differences *within* the racial groups, and, in the absence of any compelling environmental explanation for the white–Negro intelligence difference, we would be scientifically remiss not to seriously consider the genetic hypothesis.

In terms of what is already known about human evolution, about a host of other kinds of genetic racial differences, about the relative contributions of genetic and environmental factors to differences in mental abilities, and about the constancy (relative to the variability within groups) of white–Negro differences in IQ and a wide variety of other indices of cognitive development from childhood to maturity, it appears highly probable that genetic factors are involved to a substantial degree in the lower average IQ of American Negroes. So far, I have not seen a serious attempt to adduce evidence, or comprehensive argumentation based thereon, to the effect that this hypothesis is either improbable or scientifically unwarranted.

I have focused on differences between whites and Negroes in the US only for illustrative purposes and because there has been vastly more relevant research on representative samples of these populations than is true of any other racial groups. I have little doubt that other racial populations can be shown to differ behaviourally in complex ways, both in the cognitive and personality domains, and it would seem most surprising if genetic as well as cultural and environmental factors were not involved in many of these differences.

# 14  I Q and the nature–nurture controversy

*Phillip V. Tobias*

(From *Journal of Behavioural Science*, vol. 2 (1), 1974, pp. 5–24.)

## Jensen's article of 1969

The recent spate of writings on this knotty subject was triggered off by a lengthy article by Dr Arthur R. Jensen, a professor of educational psychology in the University of California at Berkeley. His treatise was entitled 'How much can we boost IQ and scholastic achievement?' As a result, in the words of Richard Lewontin (Professor of Biology at the University of Chicago), Jensen 'has surely become the most discussed and least read essayist since Karl Marx'! The article appeared in the *Harvard Educational Review* of Winter 1969. Here is a summary of Jensen's argument.

There is a large literature on Negro intelligence. It was reviewed not long ago by Shuey (1958) and by Dreger and Miller (1960 1968). The basic data in these reviews are the results of psychological tests:

(1) 'On average, Negroes test about one standard deviation (fifteen IQ points) below the average of the white population in IQ. This finding is fairly uniform across the eighty-one different tests of intellectual ability used in the studies reviewed by Shuey' (Jensen 1969, p. 81). 'When gross socioeconomic level is controlled, the average difference reduces from fifteen to about eleven IQ points'.

(2) In terms of scholastic achievement, also, 'Negroes score about one standard deviation (SD) below the average for whites and orientals, and considerably less than one SD below other disadvantaged minorities tested in the Coleman study – Puerto Rican, Mexican–American and American Indian. The one SD decrement in Negro performance is fairly constant throughout the period from grades 1 through 12' (Jensen 1969, pp. 81–2).

(3) The next 'basic fact' of Jensen is the assertion that intelligence variation has a large genetic component. This statement is based on estimates of the *heritability* of IQ test performance in middle class whites. The validity of these estimates I shall explore in a moment.

(4) Jensen hypothesizes that intelligence is determined to the extent of about 80 per cent by heredity in all human groups. Finally, Jensen puts forward the hypothesis that 'genetic factors are strongly implicated in the average Negro–white intelligence differences. The preponderance of the evidence is, in my opinion, less consistent with a strictly environmental hypothesis than with a genetic hypothesis, which, of course, does not exclude the influence of environment or its interaction with genetic factors' (Jensen 1969, p. 82). As a corollary, says Jensen, 'No one has yet produced any evidence based on a properly controlled study to show that representative samples of Negro and white children can be equalized in intellectual ability through statistical control of environment and education' (pp. 82–3).

These views of Jensen elicited support from H. J. Eysenck in London in 1971. His little book was entitled *Race, Intelligence and Education.* It led to this editorial comment in *Nature*:

The difficulty is that, true to form, Professor Eysenck has chosen to present a supposedly scientific argument without the qualifications to which any audience, but especially a lay audience, is entitled and with such a disregard for the way in which his sentences might be misquoted that his book may have the opposite effect from his expressed objective of helping to throw light on an important social problem.

In circumstances like these, is it fair of him to tell the world at large that he has arrived 'patiently and with scientific impartiality at the conclusion that the white race is inherently more intelligent (and the word is frequently used as a substitute for IQ) than the black'?

(*Nature*, 2 July 1971, Vol. 232, No. 5305)

Jensen's views were carried further by Richard Herrnstein of Harvard in *The Atlantic* in September 1971. A vigorous reaction ensued – so vigorous that the *Harvard Educational Review* is said to have refused to sell reprints of Jensen's article to anyone (including Jensen) until they could be bound together with a number of criticisms of Jensen's arguments. 'At times the arguments became so fierce that a moratorium on investigations of this kind has been suggested' (Editorial in *Nature*, 10 November 1972, p. 69).

The Society for the Psychological Study of Social Issues was moved to put out a press release on 2 May 1969, criticizing Jensen's article and stating, 'There is no direct evidence that supports the view that there is an innate difference between members of different racial groups.' The Society added that 'a more accurate understanding of the contribution of heredity to intelligence will be possible only when social conditions for all races are equal and when this situation has existed for several generations' (cited by Jensen 1971b, p. 24).

The American Anthropological Association organized a symposium at its 68th Annual Meeting, held in New Orleans in November 1969, on 'Differential Intelligence in Populations?' It has since culminated in a small book entitled *Race and Intelligence* (Brace *et al.* 1971). This contains a long exposition of his views by Jensen and four or five rebuttals by anthropologists, human geneticists and sociologists.

What was one to believe?

Undoubtedly, Jensen received ideological support and opposition too. Much of what was thundered at him or for him must have been of this nature. Perhaps, too, the support of the Nobel Laureate, William Shockley, is of this nature: for he is qualified in none of the fields which might be thought to equip a man to make informed and valid comments on the controversy – not in human genetics nor anthropology, neither educational psychology nor sociology. He is, in fact, a Professor of Engineering Science at Stanford University, and he has been co-winner of the Nobel Prize for his role in the invention of the transistor. He is, apparently, convinced that 'there is a difference in the wiring patterns' of white and black minds!

A number of critical scientific evaluations of Jensen's claims have appeared. These include studies by Biesheuvel (1972), criticizing mainly the methodology and meaning of psychological tests; by King (1971) and by Crow (1969), pointing out Jensen's misuse of heritability data culled from identical twin and other studies; by Scarr-Salapatek (1971) who has pointed up our major ignorance of most of the facts on which the issue could be decided; by Alland (1971) drawing attention to the fact that most of the tests on Negro intelligence scores (on which Jensen erected his hypotheses) fall short in methodology and quality; and by Eaves and Jinks (1972) on the statistical need for far larger test series than most workers have thus far employed.

We have to try to answer the following questions: I add here the answers to which the ensuing analysis will lead me:

1 Are IQ test scores an adequate measure of intelligence? No.
2 Does 'intelligence' or do IQ test results meet the usual scientific criteria for being regarded as an aspect of objective reality – like atoms, genes and electromagnetic fields? No.
3 Is IQ a single trait? No.
4 Is IQ an inherited trait? Not in an unqualified way.
5 Are there reported differences in average test performance among various races and social classes? Yes.
6 Have such reported differences among various races and social classes been determined validly? No.
7 Are such racial and class differences as have been found inherited? No, the evidence is lacking.
8 Will changes in educational methods, system and intensity not alter the supposed 'differences' in IQ test results? Yes, they certainly will.

### Critique of Jensen's views
First on the validity of the tests and testing:

(1) IQ's may be based on various measures of intellect. Different results are obtained, according to the test used. Hence, it must be concluded that each test assays one or more different components of 'intelligence'. For example, Biesheuvel (1972) showed that South African whites and blacks at primary school showed no difference in performance on the Porteus Maze Test, whereas, on Koh's Blocks, the white group scored significantly higher than the black. Which test should be used to construct an IQ? Similarly, Baughman and Dahlstrom (1968) found that Negro and white children at kindergarten in the rural south of the USA showed equal gains on the Primary Mental Abilities test, compared with controls not at kindergarten; whereas, on the Stanford-Binet test, Negro children showed no gains over the controls, whilst white children did. Despite these variable results with diverse tests, Jensen 'uses IQ as if it were a behavioural entity, whereas in fact it is nothing more than a statistical index

relating to tests which may measure quite different functions of behaviour' (Biesheuvel 1972).

(2) Biesheuvel (1972), John (1971) and other psychologists decry Jensen's use of the concept of *intelligence*. For Jensen treats it as a single entity (*g*) – he calls it an attribute – as definite as the concepts of energy and mass in physics. In this he adheres to one school of psychologists that includes C. Spearman and Sir Cyril Burt. For Biesheuvel and many other psychologists, notably including L. L. Thurstone and J. P. Guilford, intelligence is far more complex than that and it reflects 'many different things, differentially determined by genetics'. Thus, IQ tests assess such different components as memory, verbal skills, numerical skills, visualizing ability, systematic thought, etc. It is inconceivable on physiological grounds that all of these faculties reflect one component, '*g*', inherited as a genetic entity.

An illustrative question in one test reads: '*What's missing in this picture?*' A man is shown wearing shirt, jacket and trousers. The required answer is: 'His tie is missing.' If the subject said '*his tie*', this might measure his conformity to norms of society, and only that stratum of society where ties are normally worn. If he said '*nothing's missing*', this would lower his score regardless of whether he belongs to a stratum of society where ties are not worn!

Another example comes from the Stanford-Binet Test of 1960 for children aged 7–8: 'What's the thing for you to do when you have broken something that belongs to someone else?' The correct answer must include an offer to pay for it as well as an apology. 'Feel sorry' and 'Tell 'em I did it' are wrong answers. This clearly reflects background and attitude.

In the same way, as King (1971) has pointed out:

A cursory reading of representative tests is enough to reveal they are redolent of middle class atmosphere and values. The Wechsler Adult Intelligence Scale (WAIS) ..., one of the tests now widely used, asks in its section on comprehension: 'Why are child labour laws needed?' According to the manual of instructions for persons administering the test, one must, to obtain full credit on this question, mention at least two of the following reasons: 'health, education, general welfare, exploitation, avoid cheap labour'. If one answers by saying that child labour laws serve to keep children from competing with adults for jobs, one is given a score of zero. The whole question is considered from the point of view of an economically secure person who can afford to view children as precious social assets to be cherished and protected. Anyone whose daily struggles may cause him to evaluate children differently will score lower on the WAIS.

Thus, many different things are tested by IQ tests, some absolutely dependent on one's background, one's culture, one's upbringing; others perhaps dependent on other features. It seems clear that, of the many different things tested by IQ tests, some are not genetically determined while others may be determined in different degrees by genetic and non-genetic factors. It adds up to a picture of a series of different functional entities that these tests are assaying. To give a single overall score to the sum total is like saying that the sum of one's brain-size, one's height, one's pulse rate and the length of one's little finger equals '$g$': and then suggesting that some people have inherited different '$g$'s from other people!

(3) A most serious criticism is Jensen's disregard of several subtle environmental influences, known to participate in the complex network of causal relationships. In his 1969 paper, Jensen speaks of controlling the SES (*socioeconomic status*) of the whites and blacks being compared; but he does not attempt to control less blatant environmental influences, such as parental solicitude and warmth of interpersonal relations. These more subtle influences do not vary only with SES. They are relevant variables in their own right. Several studies have shown that these factors influence IQ test performances (Kagan 1968, Biesheuvel 1972). As a probable example, firstborns do better on tests than children born later. Presumably this is because firstborns can command greater parental time and care. I might add that a population of children that had more firstborns in it than another would be expected to have a better average test result – other things being equal. In the black population, there are on the average bigger families than in the white – this is to be expected because of socioeconomic and cultural factors. It follows that a smaller percentage of Negro children would be firstborns. Even if only one in three black children was a firstborn, as compared with one in two white children, this difference would help to depress the Negro mean test score.

Again, the different test scores and scholastic achievements of

American blacks and whites may be accounted for (at least partly) by different degrees of stimulation during infancy and early childhood.

In experimental animals, we already have some idea of *how* environmental stimulation during early postnatal life affects later neural development and behaviour (Schapiro and Vukovich 1970). It has been shown that nerve cells in the brains of stimulated animals develop complex branchings (dendrite spines) earlier than those in the brains of environmentally deprived animals. What is more, stress effects on female rats may influence the behaviour of not only their offspring but even their grand-children. Thus, the experience of the grandmother may significantly alter the behaviour of her grandchildren (Denenberg and Rosenberg 1967, Wehmer and Scales 1970). As Biesheuvel points out, the experimental evidence of this kind on man is not available; without it Jensen's generalizations are unwarranted at this stage.

However, although we may not have experimental evidence that in man *behavioural* effects of stress may still be evident two generations later, studies on at least one *structural* parameter, human birth weight, point to the probability that the depressing effect of defective nutrition persists across the generations. Thus, Hytten and Leitch (1971) summarize their survey of available evidence on birth weight, socioeconomic status and nutrition as follows:

> The evidence available, so far, though inconclusive, suggests that improvements in socioeconomic conditions and nutrition do result in an increase of mean birth weight, particularly *if the change takes place over a long enough period of time to permit a secular increase in* or nutritional revolution, even in deprived communities, would result in a rapid increase in mean birth weights . . . (p. 317) (italics mine).

Gruenwald (1968) is even more explicit. He states that there is evidence that the mother's own development during her child-hood and probably during her own foetal life affect her adequacy as a provider for the intra-uterine development of her subsequent offspring. He goes on to state: 'It is likely, if not certain, that *two or more generations are needed* for the full realization of the benefit of socioeconomic improvement' (italics mine).

If this hypothesis is valid for the very important feature of suboptimal birthweight, it is likely that the transgenerational effect applies as well to behavioural traits, such as suboptimal test performance. It becomes even more likely in the light of the animal experiments mentioned above.

Pursuing this line of thinking, we may well find that improvement of socioeconomic status must be maintained for several consecutive generations before its ameliorative effect on test performance becomes maximally evident.

(4) The evidence that, at least in rats, stress situations may have long-lasting effects, spanning even two generations, brings us to Biesheuvel's 'major and most fundamental criticism'. This is the way the environmental influences have been handled by Jensen and most earlier studies that provided his sources. When Jensen states that, for instance, socioeconomic status has been controlled, he means controlled *at the moment in time when the investigation begins*. Yet the study of a child at the age of, say, 10 years – when its family may have attained a reasonable status – tells one nothing about the position when the child was passing through its critical, formative, postnatal years, and especially the first year of postnatal life when the brain wellnigh trebles in size. Also, it tells us nothing about adverse dietary and other stress factors that may have operated in the prenatal period. Even though the present status is adequate, this does not preclude serious inadequacies having existed in previous years. Yet inadequacies earlier in the period of growth and maturation may have had serious effects on the nervous system and subsequent behavioural patterns. Thus, *earlier* environmental deficiencies may adversely affect *later* performances in schooling and in psychological tests. To control the environment *only at the time of testing* is thus liable to lull the investigator into a false conviction that he has equalized all the relevant environmental variables.

Nothing could be more misleading. Evidence has accumulated, especially in recent years, that undernutrition at critical stages during ontogeny of experimental animals may lead to permanent impairment of the brain, including diminution of brain size. This effect occurs provided the period of undernutrition begins before the cerebral cortex has reached its adult state, or, at least, during the period of maximum vulnerability of the brain to stress. If the nutritional insults are administered within this critical postnatal period, or even if it is the pregnant mother which is undernourished, the impairment to the size, structure and chemistry of the brain is not reversed by subsequent restoration of normal nutrition. These results have been obtained on pigs, rats and mice (Jackson and Stewart 1920, Dobbing and Widdowson 1965, Davison and Dobbing 1965, Dickerson and Dobbing 1967, Dickerson and McCance 1967, Dickerson and Walmsley 1967, Guthrie and Brown 1968, Zamenhof, van Marthens and Margolis 1968, Chase, Lindsley and O'Brien 1969).

What is more, longitudinal studies by Biesheuvel and his colleagues at the National Institute for Personnel Research, Johannesburg, have shown not only that rats reared on a protein-deficient diet show poor test results, abnormal brain waves and signs of focal brain damage, but also that a learning deficiency is still present in the *grandchildren* of protein-deficient grandmother rats, even though the mothers have been weaned on to a normal diet (Cowley and Griesel 1965 1966).

This effect of malnutrition is not confined to rats, mice and pigs. The studies of Engel (1956), Nelson (1959 1963), Stoch and Smythe (1973 1967), Cravioto and Robles (1965), Brown (1966), Eichenwald and Fry (1969) and Baraitser and Evans (1969) have all noted functional impairments of the human brain following early undernutrition. These workers have studied electro-encephalographic readings, psychometric testing, rate of psycho-motor development, and head circumference, to demonstrate abnormalities. Nelson (1959 1963) found that African infants, suffering from acute kwashiorkor (i.e. malignant malnutrition) showed retarded brain development which might persist, and even irreversible brain damage in some cases. Other studies elsewhere have produced comparable results, even for lesser grades of malnutrition. It is still too early to say whether the changes demonstrated are permanent; the analogy of animal experiments would suggest a permanent effect, while the study of Baraitser and Evans (1969) in the Department of Neurology at the Groote Schuur Hospital, Cape Town, would tend to support the notion that the EEG changes outlast the acute stage of the nutritional insult.

In their review, *Nutrition and Learning*, Eichenwald and Fry (1969) summarize as follows:

Observations on animals and human infants suggest that malnutrition during a critical period of early life results in short stature and may, in addition, permanently and profoundly affect the future intellectual and emotional development of the individual. In humans, it is not known whether these results may be caused by malnutrition alone or whether such intimately related factors as infection and an inadequate social and emotional environment contribute significantly to the problem. Field studies to test these hypotheses are, at best, difficult to design and to carry out; it seems likely that it will prove impossible to separate clearly the individual effects of malnutrition, infection and social environment.

In a subsequent editorial in *Science,* Abelson (1969) summarized the results of their study and of an international Conference on Malnutrition, Learning and Behaviour (Scrimshaw and Gordon 1968) as follows:

Children reared in poverty tend to do poorly on tests of intelligence. In part this is due to psychological and cultural factors. To an important extent it is a result of malnutrition early in childhood ... it seems likely that millions of young children in developing countries are experiencing some degree of retardation in learning because of inadequate nutrition, and that this phenomenon may also occur in the United States ... animal experiments suggest that (in the human infant) good nutrition during the first three years of life is particularly important.

Further follow-ups are obviously necessary to confirm the hypothesis that malnutrition at a critical age results in permanent brain damage, with permanent impairment of intellect and emotions, and of brain rhythms.

Belated after-effects of starvation and other forms of maltreatment of Second World War prisoners held in concentration camps have been investigated in Norway (Strøm 1968). Nearly twenty years later, a number of survivors showed a reduction in the size of the brain, accompanied by signs of intellectual deterioration.

On the structural side, Eayrs and Horn (1955) showed histologically that undernutrition impairs the elaboration of nerve cell processes: this may be one of the mechanisms behind both the reduced brain size and impaired brain function. Fishman, Prensky and Dodge (1969) have shown changes in the brain lipids of starved (or chronically malnourished) human beings, resembling those found in experimental animals. Not only is the total lipid reduced, but most severely affected are those classes of lipids of which myelin is composed. Here, too, we may have a clue as to the mechanism, or one of the mechanisms, responsible for the impaired function in undernourished subjects.

Although aware of these factors, Jensen is inclined to discount them. Thus he states, 'In Negro communities *where there is no evidence of poor nutrition,* the average Negro IQ is still about one SD below the white mean.' Again he writes, 'When groups of Negro children with IQ's below average have been studied for nutritional status, no signs of malnutrition have been found.' There are two serious criticisms of this rather cavalier statement that 'no signs of malnutrition have been found' at or after the time of testing. First, as Biesheuvel (1972) points out, 'Whether or not they suffered from malnutrition at a much earlier and more vulnerable stage one cannot tell from this kind of investigation, which aptly illustrates the inadequacy of the cross-sectional approach.'

Secondly, I would add this query: How certain were the investigators that undernutrition was not evident, even at the time of the tests? Its effects can be very subtle (I am not now talking about the effects of kwashiorkor, marasmus and other forms of gross malnutrition such as beri-beri, scurvy or pellagra). The effects of poor nutrition may be identifiable only biochemically; or they may be manifest physically only by a lowered growth curve, or by a longer period when adolescent girls are taller than boys of the same age, or by a depressed adult average height, or, as a most sensitive indicator, by a lesser height preponderance of adult males over adult females (i.e. lower sexual dimorphism) (Tobias 1970a 1972a). Our own studies on South African Negroes and on San (or Bushmen) have shown that such physical effects, though somewhat rarefied, nevertheless betray the presence of some environmental inadequacy, especially undernutrition. Until such environmentally-determined, apparently slight shifts of growth have been sought and two populations equalized for them, it cannot be claimed that nutritional status has been controlled even at the time of IQ testing (Tobias 1970b); nor can the assertion that it is not relevant to black-white intelligence test differences in the United States be supported.

(5) The effect of the attitude of the test-takers towards the testers has been overlooked. Yet, there is much evidence that it cannot be excluded. As long ago as 1936, H. G. Canady first asked

such questions as: 'to what extent do we know that the results of intelligence tests given by white examiners to Negro children are reliable? Can a white examiner get *"en rapport"* with Negro children? Would there be conspicuous differences between the scores obtained by a Negro examiner and a white examiner giving tests to the same group of Negro children?' He found that the race of the tester did make a difference. When tested by whites, black subjects scored, on average, five or six points below the level they reached if a black person tested them. Similar results have been obtained by a variety of later workers such as I. Katz of New York and his co-workers (1960 1963 1964 1965), Pettigrew of Harvard (1964) and P. Watson of England (1972).

It became apparent in these later studies, from the 1960s onwards, that the race of the tester was not the only factor affecting the attitude and the performance of subjects in psychological tests. Others were the examiner's age, sex and socioeconomic status, as compared with the age, sex and SES of the test-takers. In other words, the test situation came to be seen by the subject as a microcosm of society and its tensions. Thus, if the sex, age, race or class of the tester influenced a person's test performance, this was deemed so because the 'battle of the sexes', the 'generation gap', the 'race war' or 'class war' had wide-ranging effects on the group members (E. Erikson, cited by Watson 1972). Conversely, if the biological or social distance between tester and test-taker were reduced, test performance improved. For example, Gitmez (1971) showed that a test difference between a lower and a higher socioeconomic group disappeared 'when the test situation was made less threatening to the former'.

Clearly, the attitude of the child taking the test is more important than the child's response to any single question or combination of questions. If the child dislikes school, or rebels against his teacher, or is intimidated by tests, he will get a low score, no matter what his intelligence. In this way, it is possible to see how children can be unjustly channelled into programmes for the retarded on the basis of IQ tests. For these reasons, the IQ test is no longer used as a basis for channelling or tracking children in Philadelphia, New York City, Washington DC, Los Angeles and San Francisco (Mercer 1972).

Yet allowance has not been made for these factors in the studies which have provided the source material for Jensen's hypothesis. In other words, the claimed difference in IQ test scores of one SD is probably an exaggeratedly high estimate of the difference, through neglect of the need to control for these correlated variables. Yet this claimed difference of one SD is the starting point for Jensen's hypothesis – that a difference of one SD is too big ever to be caught up by educational and environmental equalization.

(6) Another aspect which Jensen tends to play down is the fairness of testing children of one group with IQ items devised by members of another. The groups referred to here may be different social classes, or different races, or as is often the case in the United States of America and in South Africa – groups differing in both racial make-up and social class.

Jane Mercer (1972) has pointed out, 'IQ tests are Anglocentric: they measure the extent to which an individual's background matches the average cultural pattern of American society', that is, of white middle class society. She found in her studies on Riverside, California, communities that the more 'Anglicized' a 'non-Anglo' child is, the better he does on the IQ test. The more so-called 'Anglo' characteristics that were present in black families, the nearer did the average IQ approach to the standard of 100 per cent. On a scale of zero to five 'Anglo' characteristics, those black families with zero to one 'Anglo' features scored 82·7 per cent, those with two 'Anglo' characteristics 88·7 per cent, three 92·8 per cent, four 95·5 per cent, and five 99·5 per cent. In other words, the average IQ of black families with five out of five 'Anglo' characteristics was virtually the same as the average IQ of the standard white community.

Despite attempts to eliminate cultural bias from intelligence tests, this has not been achieved. Indeed it appears to be a next-to-impossible task to devise a culture-free test. Yet it is true that 'a set of tests can give comparable scores only between individuals who have substantially the same cultural background' (King 1971).

Margot Smith (1973) has drawn attention to the fact that many revisions of IQ tests have been made to adapt them to the sexes, and to members of different cultural groups. But all sorts of problems of interpretation have been brought into the picture thereby. Merely changing the wording or content of a question, she found, could actually affect the difficulty of the question. Even translation into a different language could change the degree of difficulty. Thus, Sir Cyril Burt (1922) analysed sixty-five test questions. Twenty-one investigators gave the age group of children who should be expected to be able to answer correctly each one of the sixty-five questions. Only four questions were assigned to the same age group by all twenty-one investigators: the remaining questions – sixty-one out of sixty-five – varied in the age group which the twenty-one investigators expected should be able to answer the questions correctly. Again, in some versions of the Binet test, a question was placed in a particular age group when 75 per cent of children answered it correctly (e.g. Binet, Goddard, Burt); while in other versions the question was placed at that age where 66 per cent answered correctly. Thus, not only various revisions of the same test, but even the ways in which the tests are devised, proved to have been inconsistent (Smith 1973).

Margot Smith showed, too, that results varied according to the number of years which had elapsed since the test was devised, for tests grow old and cease to be useful after the lapse of time. If one then compares the performance of a group in, say, 1971, with that of the standard group at the time the revision was prepared, say in 1960, a lower score would be expected, even of the standard group. All studies on *minority* groups, she found, had been carried out at a later moment in time than the date at which the revision was devised and tested on a standard sample.

Hence test results change with changing patterns of cultural outlook, and the comparison of average test performance by various groups is thus invalidated if they are not tested at the same moment in time.

(7) Not enough is known of the degree to which environmental factors can alter test performances within an individual's lifetime. Many children, studied longitudinally, have been shown to change IQ scores by twenty points or more from childhood to adulthood. Eysenck (1971), though convinced that US blacks are genetically inferior, is none the less optimistic about the potential effects of radical environmental changes on Negro IQ scores. He stresses that very large IQ gains are produced by intensive one-to-one tutoring of black urban children with low IQ mothers. In intensive programmes of this sort, there are large environmental changes and large IQ gains result. In contrast, in other programmes with insignificant environmental improvements, only small IQ changes are obtained. He observes, correctly, that large IQ changes may be produced by creating appropriate, radically different environments never before encountered by those genotypes. Surprisingly, however, he expects that new environments,

such as that provided by intensive tutoring, will not affect the black–white IQ differential. Yet, it is a fact that many middle class white children *already* have learning environments similar to those provided by tutors for the urban black children (Scarr-Salapatek 1971). The inference seems justified that such steps alone are likely to reduce markedly the black–white differential. Clearly we need much more research on the possibilities of such new environments, before we abandon the idea that environment is a major determinant if IQ scores.

## The determination of heritability by twin studies

A large part of Jensen's argument hinges on his claim that IQ variance has a heritability of 0·8 (or 80 per cent). That is, he accepts that 80 per cent of the variance of the IQ of whites is contributed by genetic factors – and only 20 per cent by non-genetic or environmental factors. He then assumes the figure 80 per cent to be true also for Negroes, as well as for white–black IQ test score differences. From this, he argues that, even if all environmental influences were equalized, this 'could not possibly raise the mean for the blacks by one standard deviation' (the average difference between black and white test scores). And he goes on to urge the abandonment of Operation Head Start and other US programmes of compensatory education.

We have already questioned the meaning and the validity of the supposed one standard deviation difference. Clearly, a great deal hinges on the correctness of the heritability figure of 0·8 and of the way in which Jensen uses the figure. On both counts, Jensen may be criticized. The figure of 0·8 has been derived, in the main, from studies on separated identical twins (as well as on adopted children); we shall have to take a look at the twin method in human genetics.

One way to elucidate which traits are hereditary in a given population exposed to a particular environment is by researches on identical twins. Fraternal or two-egg twins are no more alike genetically than any two children of the same parents. Identical or one-egg twins start life with the same (or virtually identical) genes as each other. Whatever differences arise later may be attributed almost entirely to the interaction of varying environments with the same basic set of genes. If we study those things in which identical twins agree and those in which they disagree, we might arrive at a first sorting of traits. For example, if one member of a pair of identical twins develops diabetes, there is a strong likelihood that the co-twin also will develop diabetes, and we should be inclined to attribute this to a common genetic predisposition to diabetes. If both show blue eyes and a fair skin, we should again reasonably ascribe these traits to genetic causes. If, on the other hand, one twin gets measles, while his co-twin escapes it, we might tend to think that this was an environmental effect. So we categorize traits into those for which both members of a twin pair show the feature or are *concordant*, and those in which they disagree or are *discordant*.

Does it follow that all traits for which identical twins are concordant are genetic, and all traits for which they are discordant are non-genetic? At first blush, this may seem so, but a moment's thought would show the error of such thinking. Two members of a twin-pair may both develop chicken-pox; this, however, does not prove that chicken-pox is hereditary! If the twins share the same bedroom, or even the same house, it is very likely that, if one twin picks up the infection, the co-twin would acquire it too. Or again, if they are brought up in the same home, it is likely that both will embrace the same religion. But their concordance in religion does not mean that the choice of a particular faith is governed by genes! In this instance, it is clearly the effect of a common home environment acting on both twins.

In other words, if identical twins grow up in a virtually identical environment, their concordance for some traits cannot automatically be interpreted as proving that those traits are genetic.

If, however, identical twins are separated soon after birth and brought up in two different households, i.e. in two different environments, the amount of agreement between them *for any trait* may be accepted as indicating the strength of the genetical determinants of that trait in this pair of twins. And the amount of difference between them *for any trait* may be taken to reveal the degree to which the two different environments have modified the expression of that trait in the two twins. Suppose, for instance, one member of a twin pair is given out for adoption immediately after birth and is adopted into a childless home. Suppose now the two twins are examined at 20 years of age, that is, after each has been exposed for twenty years to an environment different from the other. Any difference in physical traits between them then would suggest the effects of the two different environments. Thus, if one were taller than the other, this would seem to indicate an environmental effect. But the position is not quite so simple.

Obviously, small differences can arise between identical twins, even when reared in the same home. Hence, what we must do is to compare the amount of difference between twins reared apart with the amount of difference between twins reared together. To do this, we should need, say, 100 pairs of identical twins reared together, and 100 pairs reared apart. If we studied a trait which could be measured, e.g. stature (height), we could compare the average difference in height between co-twins reared apart, with the average difference between co-twins reared together. Any increase of the average inter-twin variation in those reared apart could fairly be laid at the door of environmental differences.

Thus, by studying many traits in large samples of twins reared apart and together, we could arrive at an idea of which traits in these samples of twins were in general more dependent upon genes and less modifiable by the environment; which were about equally determined by both sets of factors; and which were less genetic and more environmental.

Four major studies of separated identical twins have been carried out, one in Chicago by Newman *et al.* (1937), two in England by Shields (1962) and by Burt (1966), and one in Denmark by Juel-Nielsen (1965). These studies have attempted to determine the *heritability* of physical and mental features. By the heritability of a trait is meant the *proportion of that trait's total variation in a population attributable to genetic differences* (as distinct from that proportion attributable to non-genetic or environmental differences). When we apply this idea to a 'population' of identical twins reared apart, clearly the heritability will be greater, the less difference there is between the two sets of environments experienced by the co-twins reared apart. Thus, if an adopted twin is brought up in a home differing very little – in SES, educational level, diet, emphasis on exercise – from the home of the co-twin, clearly the environmental difference will contribute little *extra* inter-twin variability to any trait under consideration. In such a case, environmentally determined variance would be slight and virtually all variance between the twins would be genetic: heritability would be high. On the other hand, if the average gap between the two sets of home environments was great, it is to be expected that the environmental effects would contribute much more of the inter-twin variation, while the genetic contribution to the inter-twin variation in this case would be appreciably smaller: heritability would be low. In this way, depending on the average amount of environmental difference, *the same trait* could yield different heritability values in different studies.

## Heritability and intelligence

In the light of this, it is interesting to see that the correlation in IQ scores between co-twins reared apart varied in the four studies mentioned – from 0·62 (Juel-Nielson 1965) to 0·86 (Burt 1966). In the studies of Burt, only a very slight environmental influence is evident; the heritability contribution is high. In the study of Juel-Nielsen, the correlation is lower, and the environmental contribution greater.

Hence, for the trait or, better, the complex of traits, measured as IQ, the heritability, determined from twin studies, is not constant. It varies inversely with the diversity of the environments: the more different the environments, the lower the heritability; the more alike the environments, the higher the heritability. In the four studies cited, heritability estimates vary among themselves – so environmental contrasts were not constant. Nevertheless, all four heritability estimates were high, the average being 0·73 or 73·0 per cent. Now the four studies used middle class, white Americans, Britons and Danes; undoubtedly, the contrasts between the environments to which the two twins of each pair were exposed were not great. Kamin (1973) has pointed out just how little the two environments differed in some instances, by the following illustrative examples from the study of Shields (1962):

*Benjamin and Ronald*, separated at 9 months: 'Both brought up in the same fruit-growing village, Ben by the parents, Ron by the grandmother ... They were at school together.... They have continued to live in the same village.'
*Jessie and Winifred*, separated at 3 months: 'Brought up within a few hundred yards of each other.... Told they were twins after girls discovered it for themselves, having gravitated to each other at school at the age of 5.... They play together quite a lot ... Jessie often goes to tea with Winifred.... They were never apart, wanted to sit at the same desk....'
*Bertram and Christopher*, separated at birth: 'The paternal aunts decided to take one twin each and they have brought them up amicably, living next door to one another in the same Midlands

colliery village.... They are constantly in and out of each other's houses.'

They may be rather extreme examples, but they do show the kinds of case that were included under the category of 'identical twins reared apart'! (Kamin 1973).

Certainly, these environmental disparities are extremely small compared with the total range of environments to which modern *Homo sapiens* is exposed – from Eskimo igloos to Bushman shelters.

Hence, it is reasonable to infer that, because the backgrounds are relatively homogeneous in these studies, the environment has contributed little extra variation. Under these circumstances, the four studies give an inflated estimate of the heritability. Then, too, it is the heritability only for the white middle class population of the USA, Britain and Denmark, and it is a within-group estimate. If the two members of each pair of identical twins had been reared in markedly contrasting situations, environmental effects would have been greater – and heritability lower.

For example, in the classical 1937 study by Newman, Freeman and Holzinger, eight IQ points was the average difference in IQ between members of identical twin pairs who were reared apart. While the mean within-pair differences in IQ was eight the differences in individual pairs of twins ranged from one to twenty-four points. This shows just how much difference in IQ performance environmental differences can bring about in genetically identical pairs of individuals. Similar and even more striking results were obtained by Shields (1962) in his study of identical twins reared apart.

As Crow (1969) said, 'it is obvious from looking at the data on identical twins that individuals with exactly the same genetic constitution can differ widely in the phenotypic trait we measure with IQ tests and label intelligence'.

In spite of this fundamental consideration, Jensen has assumed that the middle class white figure of 0·73 (or 0·8 as used by him) can be applied automatically to the members of any other race or social class. Thus, he applies the same figure for within-group variance to disadvantaged white children, and to black children – many, if not most of whom were disadvantaged. Until recently, there were no factual data on the heritability of IQ scores in black peoples or in any racial groups other than the white or Caucasoid.

Since our knowledge of within-group variance is so slight, it follows that we know even less about the heritability of *between-group* variance of intelligence, e.g. the black-white differences collated by Shuey (1958). And, as Thoday (1969) pointed out: 'There is no warrant for equating within-group heritabilities and between-group heritabilities'. This is not to say that we can deny the participation of a hereditary component in between-group comparisons: we do not have evidence to exclude it, on the one hand, nor to assess its possible magnitude on the other. The difficulty is aggravated – so that the problem becomes at present virtually inaccessible to scientific investigation – by the fact that, to a very considerable extent, the difference in environment between blacks and whites in the United States is determined by whether the individual is classified as black or white. This makes it practically impossible to randomize the environment (King 1971). It is even more difficult, indeed absolutely out of the question, to do so in South Africa, where legislation, as well as common custom, *lays down* differences in environment for various ethnic groups.

Talking of identical twin studies, some wit suggested that the only way one could prove that there are major racial (i.e. genetic) differences in intelligence would be to dye one member of a pair of white identical twins black and adopt it out to a Negro family, while the co-twin is reared by a white family! How much difference would it make to their IQs? Another jokingly suggested that we should find pairs of identical twins in which one member of each pair is Negro and one is white, separate them at birth and rear them in Negro and white families, and see how their IQ differences compare with those found for twins where both are of the same race! Conceptually sound, maybe, but the one is unfeasible, the other impossible (Jensen 1971b).

## IQ, environment and genes

A recent study by Scarr-Salapatek (1971) has thrown some light on the matter. She has started out from this standpoint: 'Dislike of a genetic hypothesis to account for racial differences in mean IQ scores does not equal disproof of that hypothesis. Evidence for genetic or environmental hypotheses must come from a critical examination of both explanations, with data that support one.' She continues: 'There is no reason to assume that behaviours measured in one population will show the same proportion of genetic and environmental variances when measured in a second population whose distributions of genetic or environmental characteristics, or both, differ in any way from those of the first population.'

She has carefully analysed both the so-called *environmental disadvantage hypothesis* and the *genetic differences hypothesis*. Then she records the results of a new study on about 1000 pairs of twins from both black and white families in Philadelphia. Her results on aptitude scores confirmed the black-white difference in mean scores of about one standard deviation. She found the heritability estimates varied greatly with social class. Among disadvantaged children, [1] whether black or white, the proportion of genetic variance (heritability) was low, among advantaged children, heritability was generally higher, and more so for verbal aptitude tests than for non-verbal tests. In other words,

A closer look at children reared under different conditions shows that the percentage of genetic variance (i.e. the heritability – PVT) and the mean test scores are very much a function of the rearing conditions of the populations. A first look at the black population suggests that genetic variability is important in advantaged groups but much less important in the disadvantaged. Since most blacks are socially disadvantaged, the proportion of genetic variance in the aptitude scores of black children is considerably less than that of the white children ...

Thus, Scarr-Salapatek comes down squarely in favour of an environmental hypothesis, rather than Jensen's genetic hypothesis; though it has been questioned whether her sample of 1000 twins was large enough for her conclusions to be statistically foolproof (Eaves and Jinks 1972). Incidentally, she qualifies her viewpoint by an *additional cultural hypothesis*. According to this view, the environmental factors *do not act in the same way* on black as on white children: home experience has a different *relevance* to scholastic aptitudes and achievement (Coleman *et al.* 1966, Irvine 1969). In black populations of Central and East Africa, the quality of school is the only environmental variable which significantly and consistently influences ability and attainment tests; there is no such influence of socioeconomic status, family size or family position. These extra-scholastic or home influences are much more important in European and in American white children. Scarr-Salapatek suggests that for the black child in Philadelphia, the relevance of home experience is surely greater than it is for the 'tribal African'. But, 'one may question the equivalence of black and white cultural environments in their support for the development of scholastic aptitudes ... the black child learns a different, not a deficient set of language rules, and he may learn a different style of thought. The transfer of training from home to school performance is probably less direct for black children than for white children.'

She thus accounts for the differences in black-white mean test scores by environmental differences (those between the disadvantaged and the advantaged). She tempers this view by a concept of cultural differences, according to which the rearing environments of young black children are less conducive to the development of scholastic abilities, and those of young white children more conducive. Perhaps the so-called cultural differences are simply another facet of the contrast between the deprived and advantaged home backgrounds.

It is important to realize that heritability is a relative measure: an expression of what proportion of the variability of a trait in a particular population is genetic variability as compared with what proportion is environmental. Because of the way we measure heritability – using the twin-method or adopted-child method – its value varies according to the diversity of the environments. Thus, its value for one group with very uniform middle class environments will not tell us anything about its value in another group spread over a greater range of environments. Hence, heritability is a relative measure and it tells us nothing about the degree of genetic determination of the traits measured by IQ in mankind at large, let alone about the differences between races.

As Alland (1971) has put it, 'Jensen has taken a fairly safe hypothesis – that intelligence is heritable – and forced it to carry the burden of a second argument for which there is still little evidence: that black and white performance on intelligence tests is determined primarily by genes.'

**Race and intelligence: summation**
While one cannot exclude the possibility that there is an appreciable genetic factor in the variability of some of the component traits assessed by various IQ tests, at the same time it is scientifically unjustified to assign to such possible genetic factors a precise value (like 0·8), to apply such value to all within-group

and between-group comparisons, and then to base inferences on such value. Hence, it is not valid to infer that, 'because of such high heritability', members of a disadvantaged population cannot hope to elevate their new mean test scores by as much as one SD by the elimination of all environmental differentials. This is a fundamental flaw in Jensen's use of the genetical concept of heritability: this glaring misuse of the concept, *on its own*, invalidates the methodology of Jensen and, therefore, his hypothesis. When to this objection are added all the other criticisms mentioned, there is no option on presently available evidence but to reject Jensen's hypothesis. I find myself in agreement with King (1971) when he states: 'The fact is that we have no evidence on the question of the inheritance of difference in intelligence between races, and we are not likely to get any until we discover means for greatly improving our techniques of investigation.' Biesheuvel (1972) has spelt out a similar conclusion:

> I believe that Jensen has failed to establish a valid case in support of the hypothesis he has put forward. It is not possible to test this hypothesis by means of the type of cross-section or cross-cultural group comparisons, or the correlational studies that are usually undertaken. I consider that the establishment of adequate controls for representative samples is impracticable. Instead it is recommended that the problem be attacked by means of intra-group longitudinal studies in which we attempt to determine the limits of modifiability of behaviour in relation to a number of factors, varied both singly and in various combinations. This type of research is tedious, costly, and unlikely to yield conclusive answers for some time to come.
>
> Either one must accept these conditions, or leave the problem alone. Meanwhile, the scientific evidence available at this stage on the genetic determination of race differences does not justify categorical statements which could have a major impact both on public policy and on race relations. (p. 93)

The conclusion of two distinguished geneticists, W. F. Bodmer and L. L. Cavalli-Sforza (1970) is similar: 'We do not by any means exclude the possibility that there could be a genetic component in the mean difference in IQ between races. We simply maintain that currently available data are inadequate to resolve this question in either direction.' (p. 29).

In the words of Sandra Scarr-Salapatek (1971): 'To assert, despite the absence of evidence, and in the present social climate, that a particular race is genetically disfavoured in intelligence is to scream 'FIRE! ... I think' in a crowded theatre. Given that so little is known, further scientific study seems far more justifiable than public speculations.'

## Conclusions

From all the evidence reviewed here, it is seriously doubted that IQ tests are an adequate measure of intelligence, or that they measure up to the usual standards of rigour, consistency and objectivity demanded by scientists.

The evidence flatly contradicts the idea that what the various IQ tests measure constitutes a single trait, in the sense that eye colour, skin pigmentation or even stature are single traits. Instead, such tests appear to assay a number of different items of behaviour, some of which are manifestly non-genetic in causation, and others of which are of as yet unknown causation, with the possibility that genetic factors and non-genetic may play a variable role in the determination of some of these items. There is no adequate basis for the claim that the deficit reported from IQ tests on blacks reflects predominantly a difference in genetically determined intelligence. It must after all be admitted that science has not yet validly demonstrated any *genetically determined* differences in kinds of nervous systems, patterns of behaviour or level of attainment among the races.

Hence, it is misleading and totally unjustified to apply genetic analysis to the IQ score – as though, by expressing so hetero-geneous a pot-pourri of behavioural items as a single score, one had thereby converted them collectively into a single trait.

Moreover, the evidence strongly indicates that the deficit (of one standard deviation) reported in IQ tests on blacks is not a valid or fair estimate. It is exaggeratedly high, for a number of reasons which have been reviewed here and which are connected with sampling and testing procedures, the timing of the tests, cross-cultural, cross-racial and cross-SES biases, and other factors.

This survey finds that the balance of evidence presently available must lead to the rejection of the claim that blacks will never catch up their reported IQ deficit; while some presently available evidence even suggests that the deficit can be largely wiped out by more adequate cultural and socioeconomic controls and by intensive compensatory tutoring.

Hence, because of our profound ignorance and the inadequacy of our evidence, we dare not recommend the abandonment of all compensatory education. Rather, we ought to try to produce drastically improved and intensified education, over several successive generations, and not only of schoolchildren, but of their parents who provide – or should provide – a large part of their children's extra-scholastic education. The results may be amazing.

This is, I believe, the correct attitude to take at the present, based upon a sober, balanced evaluation of the facts and an appreciation of our ignorance. I cannot deny that races may differ genetically in psychological traits; but then I cannot assert it either (Tobias 1972b). As Biesheuvel (1952), Schwidetzky (1967) and others have pointed out, there may be differences not in the level that may be attained, but in aptitude – 'bent' towards particular abilities, rather than of general intellectual capacity. 'There may be a real difference in the mean level that can be achieved; or there may be no difference at all, neither qualitative nor quantitative, given adequate opportunities for development' (Biesheuvel 1952).

## Acknowledgement
I am indebted to Kay Copley and Carole Orkin for their invaluable assistance.

## Notes
1 'Disadvantaged' as used here connotes all of the biological and social deficits associated with poverty, regardless of race.

# 15 The effects of preschool education: some American and British evidence

## G. Smith and T. James

(From *Oxford Review of Education*, vol. 1, No. 3, 1975, pp. 223–40.)

Social research findings rarely point unswervingly to clear policy conclusions. The recent history of preschool research and practice well illustrates the twisting, turning relationship. In the 1960s, as policy interest focused on preschool education, the lack of any research on its effects, particularly in poverty areas, served as an indictment of the way social research ignored important practical questions. [1] But a decade later, having inherited the confusing mass of American material from Head Start and other projects, are we any better placed to offer clear guidance on preschool policy?

It is true that at times preschool research has appeared to endorse a simple straightforward message. Bloom's original findings about the amount of variance in adult IQ scores that could be predicted at an early age were rapidly interpreted into unfounded statements about the 'amount of intelligence' formed at any age point (Bloom 1964) [2] – a 'truth' that finds its way, for example, with spurious precision, into the Ladybird *Learning with Mother* series – 'Almost half of your child's intelligence will be decided by four and a half.' Enrolling children in preschool groups seemed the answer to almost every social or educational problem. The reaction, when these problems were not solved, was equally dogmatic: 'Head Start has failed' or the even more general 'education cannot compensate for society'. But these positions have hardly been taken up before they have been challenged by further research and experiment. The problem in the face of conflicting results is to draw out any coherent patterns and establish the implications for preschool policy.

In a recent review of preschool research in Great Britain, Barbara Tizard (1974) has bravely attempted to summarize some of the conclusions. She suggests that as a result of American and

British research there are now generally agreed answers to six of 'the simpler questions about the influence of nursery school attendance on children's achievements'. These are:

1 That conventional playgroup or nursery experience without special programmes has no significant effect on test scores or primary school achievement.

2 Nor does it have any long term impact on general adjustment or classroom behaviour in the primary school.

3 Special programmes, if supported by staff, can produce significant increases in test score.

4 Children of different social backgrounds benefit equally from these programmes.

5 Gains occur rapidly and may be as much the result of 'test-taking familiarity' as a generalized increase in cognitive skills.

6 Without follow-up programmes, preschool gains tend to 'wash out'.

Taken together, these answers incline Barbara Tizard to a pessimistic conclusion about the effects of preschool: 'In so far, then, as the expansion of early schooling is seen as a way of avoiding later school failure or of closing the social class gap in achievement, we already know it to be doomed to failure'. Yet though each of these answers may represent the best overall summary of research, in almost every case there is alternative and more complex evidence. As a result the conclusion may turn out to be too pessimistic.

In presenting selected findings from preschool experiments from the West Riding Educational Priority Area project (Smith 1975, James 1975), our purpose is to suggest some of these alternative possibilities, and sketch in a framework that may help to explain some of the conflicting results. Rather than present these findings in a vacuum, we have prefaced them with an interpretation of American preschool studies. Though this is now a well-reviewed field, we make no apologies for attempting to make our own reading clear.

We have adopted a semi-historical approach to the evidence for two main reasons. First, the preschool debate has undergone several sudden changes of direction. The conclusions drawn at any one time are heavily dependent on the expectations generated and the questions asked; they have to be placed in a historical context. Second, the usual time lag of a year or more before American evidence makes its full impact in this country has increased the confusion, with different groups responding to different phases of the debate – some still buoyed up with the early optimism of studies such as Hunt (1961) and Bloom (1964), others depressed by the long string of negative findings, particularly the massive Westinghouse study of Head Start (Cicirelli *et al.* 1969). These changes of direction present a major problem in any review; is the purpose to cool optimism or reduce pessimism about the effects of preschool work? In the final analysis, the answer is bound to be relative: a significant improvement for one group will for another seem negligible. Our own reading of the most recent evidence suggests a move away from the most pessimistic phase, particularly now that the debate has shifted away from a simple success/failure mentality with the central question 'is preschool effective?', to a more detailed examination of the effects of different forms of preschool experience.

## The American evidence
Recent developments in American preschool research can be grouped into three distinct phases; naturally these overlap, as the change from one phase to another often represents intervention by a new group, rather than any change of mind by the original set. The early psychological and educational emphasis in preschooling for the disadvantaged was later joined – and undermined – by a more sociological approach, which threw doubt on the value of short run educational gains when little else in the child's environment had been altered. Anyone now entering the preschool debate has to argue on many fronts, before an audience of different research interest groups. This may be a necessary and welcome feature of policy related research – rescuing questions from the control of narrow sectarian interests – but inevitably it becomes harder to find answers satisfactory to all.

### The first phase
The first phase is almost too well known. The combination of Hunt's summary of animal development studies, and Bloom's reanalysis of longitudinal data on child development focused attention on the possibility of rapidly accelerating development during the preschool years, particularly for children whose early experiences were restricted. Both studies reinforced a 'once and for all' mentality, where something done in the early years had independent lasting qualities.

Translation into practice was simple enough. In the first phase the race was to discover the form of preschool curriculum that would most rapidly accelerate early learning. The main approach was the experimental preschool project often set up under near laboratory conditions on college campus and a rash of these followed the growth of funds available for such experiments. The work of Gray and Klaus, Deutsch, Weikart, Nimnicht, Karnes, Blank or Bereiter and Engelmann, to name some of the best known, has been extensively reviewed elsewhere. At first the results seemed to point conclusively to a clear answer. Weikart's (1967) analysis of different approaches, grouping them under 'traditional', 'structured' or 'task-oriented' methods, indicated that the more highly directed 'task-oriented' approach – the best known example being the Bereiter and Engelmann (1966) programme – the larger the gain achieved. So whether one approved of such directive methods, or supported procedures such as 'operant conditioning', the answer seemed to be that 'they worked'. Given the high confidence that rapid and dramatic gains would occur, it was perhaps natural that the most frequent tests used to evaluate the effects of these experiments were global IQ measures, rather than any more fine grain assessment.

The conclusion about structured programmes was already well established as preliminary results began to come in from the national Head Start programme, launched on the same wave of interest in preschool education as an eight week summer programme in 1965, and on a full year basis shortly afterwards. Though initial results from Head Start were uneven, they did not generally show the large gains achieved in the experimental projects. Head Start was, of course, set up on a national scale using different local agencies, and was never a uniform programme. Within its broad mixture of education, social development, nutrition and health care, there was little emphasis on structured curriculum, and indeed evidence that this was resisted by workers recruited locally who were committed to a more traditional form of preschooling.

The lack of a highly structured curriculum was an obvious reason at hand to explain the relatively poor performance of Head Start centres; and the obvious solution – widely canvassed by researchers – was for them to adopt the structured curriculum methods developed at experimental centres. One result of this pressure can be seen in the development of the 'Follow Through' programme and in the later Head Start 'planned variation' study, described in more detail below. Both of these experiments aimed to test out different curriculum approaches to provide comparative information on their effectiveness. But by the time these studies were under way, the position had basically changed. The belief that the more structured the curriculum, the more substantial the gain, was shown to be at best a partial and short-run truth.

Critics anxious to reassert the genetic basis for differences in measured ability between different social and ethnic groups, and to counter the 'environmental' bias of this optimistic phase, have pointed to the essentially 'plastic' conception of human nature underlying much of this preschool intervention. Basic changes could be achieved, if only the magic curriculum key could be found. But it is also an asocial view, treating 'environment' as a one-off event and ignoring the persistent social pressures and constraints on child, family and school.

### The second phase

The second phase is marked by two competing developments. On the one hand evaluation and research accumulated, either showing the long-term ineffectiveness of preschool education, or, worse, casting doubt on whether it could ever be effective; on the other, there were increasingly elaborate attempts to explain away the lack of clear cut success, and devise programmes to avoid earlier failures.

It is not our intention to review these follow-up studies in detail. However, what quickly became an established truth was that preschool gains soon 'washed out'. In practice this was a variable effect. Some programmes, like that at Nashville (Gray and Klaus 1969), were able to show differences between experimental and control groups in a seven year study, though the experimental group had by then fallen below its original starting point in terms of standardized score. In other cases, the most rapid gains were quickly followed by equally rapid decline (Bereiter 1969). Though the pattern of results could have been interpreted in many ways, the simple conclusion drawn was that preschool did not work. For it had not provided the promised 'once and for all' boost in development.

The national Westinghouse study of Head Start (Cicirelli *et al.* 1969), which investigated follow-up effects up to three years after preschool, came as the final blow. Though its findings were not out of line with other studies of Head Start, the extent of its coverage, follow-up and testing procedures left little room to argue that there were somewhere substantial effects which had escaped notice. Critics could point to the way a national study inevitably grossed up the good, bad and indifferent to produce an overall verdict of no change (Smith and Bissell 1970); and there is indeed evidence in Westinghouse that centres in certain areas were more effective than others; but again this hardly amounts to the dramatic gains anticipated. In a different climate, more might have been made of these regional variations or the residual effects on cognitive development still detectable three years after the programme. But expectations had been high; as Bereiter (1972) pointed out, we were 'asking the doctor for a pill [children] can take when they are ten that will prevent them getting fat when they are fifty'.

Follow-up studies undermined the claim that early education would have a once and for all effect; research by Weikart (1972) undermined the belief that one type of curriculum was necessarily more effective than another. In a series of replicated experiments, comparing three different approaches, described as 'cognitively oriented', 'language training' and 'unit based', Weikart showed that there was little difference in the results. Where the unit-based approach, closest to the traditional form, appeared to become less effective, Weikart attributed this to the decline in staff morale, as the other more experimental programmes received greater attention, more visitors and prestige. In this experiment, all three treatments were directly under the project's control, allowing equal staff time, planning and enthusiasm. Weikart concluded that what he calls the 'staff model' was more important than curriculum: this meant time and organization to plan, adequate supervision and staff morale. Curriculum is 'for teachers', to focus their energy systematically, provide a framework for deciding what is relevant to teach, and furnish criteria to allow others to judge the programme's effectiveness.

Bereiter (1972) gives one explanation why more structured programmes may in previous experiments have produced larger gains than traditional approaches. He suggests that the traditional approach is distinguished most 'on the basis of things that are not done'. 'Teachers in the traditional programmes are not so much distinguished by differences in relative frequency of different kinds of teaching acts . . . as by the generally low frequency of

teaching acts of any kind.' In Weikart's study it seems likely that
the systematic development of the traditional 'unit-based' approach would have avoided this weakness.

One response to this depressing set of results was to explore
alternative ways of making early education more effective. Several
of these approaches were already being tested out, but now
received far more attention. The first was to attempt intervention
at an even earlier age, as preschool programmes for 3-year-olds
were already 'remedial' in character. The strategy was for home
visiting programmes with young children and their parents,
which began almost at birth. Again there was the same pattern of
successful experiment; reviews suggest that home visiting programmes have so far shown less 'wash out' effects, though there
have not as yet been many follow-up studies to the point in school
where rapid decline seems most likely to occur. Bronfenbrenner
(1972) suggests that the success of home visiting lies in the
combination of an early start – before three is vital – and the
involvement of a sustaining agent – the mother. Parental
involvement is both a 'catalyst' and a 'fixative'. Certainly home
visiting allows direct contact between teacher and parents, often
difficult in a conventional preschool group, and concentrates
individual attention on the child. However, the example of
preschool curriculum suggests caution before accepting that there
are any forms of early education that will ultimately proof the child
against later experience.

A further response has been the development of follow-up
programmes at school level. As Gray and Klaus (1966) point out,
preschool programmes 'can only provide a basis for future
progress in schools and homes that can build on this early
intervention'. Their position is supported by the findings of Wolff
and Stein (1967). In a study of Head Start centres in New York
City, it was found that where children from Head Start had joined
a school class with a sympathetic teacher, gains were maintained.
When the teachers were poor, children from Head Start tended to
regress to lower levels than when they began preschool. The
authors conclude, 'Head Start advantages can be maintained only
if the level of teaching and curriculum in the kindergarten are
strong. It implies the opposite as well – that more damage is done
to the child who looks forward eagerly to an educational
programme he has learned to enjoy than to a child who has had no
previous knowledge of what to expect, if the later school
experience is poor'.

So far, however, the effects of long-term follow-up intervention have been mixed, not least because of the technical problems of following children over time. Bronfenbrenner points to the lack of results from sustained intervention where there have been two or more years of special programmes; in the Deutsch study in New York, there were no significant differences between the experimental and control group after five years continuous intervention. However, he also reports more optimistic findings from the 'Follow Through' programme. Originally intended as the national follow-up programme for children who had experienced Head Start, this programme was initially run on a large-scale experimental basis. Initial results showed that children in Follow Through programmes made larger gains in achievement during the school year than the comparison sample. Other encouraging results were that children from families below the OEO poverty line tended to improve more rapidly than others, and those who had experienced Head Start before joining 'Follow Through' also made higher gains. These are preliminary results, though they are at variance with some of the earlier studies. Bronfenbrenner tentatively attributes this difference to the broad nature of Head Start and Follow Through, which provide not only education, but health care and family services. He implies, too, that different forms of curriculum are needed at different age points, given the general nature of educational and social disadvantage; a highly structured cognitive programme could not be adequate on its own.

The second phase, then, was one of accumulating evidence against the more optimistic claims for preschool education. Its supporters were ground on the one hand between the theories of Jensen, reasserting that group differences in measured ability were genetically rather than environmentally determined and therefore not open to remedial intervention at least of a conventional kind – and on the other by the work of Jencks and his colleagues, which suggested that even if preschool education could achieve large gains, this would have little effect on later social and economic advantage. The fault lay not in the socialization patterns of the poor and their educational achievement, but in the opportunity structure of society. Yet the second phase also encouraged the development of alternative strategies such as home visiting, a search for alternative objectives in addition to that of simple cognitive development, and a better understanding of the limitations of preschool intervention. The result has been a more realistic assessment of its potential in the current, third phase.

### The third phase

The third phase again divides into two separate strands. Few now believe that preschool could by itself produce the long-term changes needed if the gap in attainment between different social groups is to be closed. The idea of 'the magic years' is virtually dead; yet it refuses to lie down. The arguments for early intervention on both developmental and social grounds remain powerful; as Bruner (1975) writes, 'the staggering rate at which the preschool child acquires skills, expectancies and notions about the world and about people; the degree to which culturally specialized attitudes shape the care of children during these years – these are the impressive matters that lend concreteness to the official manifestos about the early years'. And however we seek to explain away the gains made in experimental programmes, there have indeed been gains on a substantial scale, maintained for a period not so far achieved at other levels of education. That they were temporary does not necessarily prove them an illusion.

There is thus still life in the search for an effective form of early education. Though no one component – home visiting, parental involvement, special curriculum or school follow-up – seems on its own to make the decisive impact, what would be the result if they were combined in a long-term 'sequential' programme? Bronfenbrenner reports the development of Heber's study along these lines in Milwaukee. This began by providing day-long intensive programmes for mothers and children initially on a one to one basis, with impressive early results. One obvious problem is the prohibitive cost of implementing such an approach on any scale, as the attempt is to provide almost total institutional care as well as individual attention. Yet even this intensely high quality early experience may lead to the same pattern of declining results, if later school experience and the wider environment are not similarly improved. Bronfenbrenner draws attention to the priority for what he calls 'ecological intervention' – a more acceptable term perhaps for the politically more controversial social and economic changes required if the 'ecology' of poverty areas is to be significantly altered. Perhaps, in the long run, this would prove an easier and less costly alternative to the increasingly elaborate attempts to compensate the child for – or even remove him altogether from – his environment.

At the other end of the scale there has been more detailed examination of the effects of different forms of preschool experience. Findings here throw more light on earlier conflicting results, and suggest that there may be real dilemmas in selecting one form of curriculum rather than another. Bissell (1973) discusses early results from the Head Start 'planned variation' study and from the similar experiment in project Follow Through. In the first, eight different curriculum approaches were tried, and in the second fourteen; these ranged from intensive academic teaching programmes to approaches based on the discovery methods of English primary schools. Though the research is plagued by the difficulties of comparing several programmes with different goals and by major technical weaknesses, it has produced tentative evidence that though well-implemented programmes may have similar results on some measures, on others there are specific effects related to programme content. First research established that programmes in operation actually were different from one another. On overall measures of cognitive development, children taking part in the Head Start or Follow Through experiment tended to make larger gains than a comparison group which had not, and, in line with other studies, the Head Start 'planned variation' results showed that the more structured programmes were more effective in raising cognitive development. However, Bissell adds details of more specific effects. In an analysis of children's response patterns on the Stanford-Binet test, it emerged that children in the academic programmes were more likely to give 'passive' responses when they did not know the answer. As Bissell argues, these changes suggest that the children have learned what a question is and what an 'appropriate' answer is, and that they have learned to focus their attention on the the components of school-like tasks. Bronfenbrenner, however, queries whether such skills will be as applicable at higher levels of education, where independent initiative, rather than appropriate response patterns, may be called for.

In the Follow Through study, though children in the academic programmes again made larger gains than those following discovery methods, the favourable shift in attitudes towards school was more marked in the latter. Also there was 'a strong association between gains in achievement and positive shifts in attitude towards school', in the discovery and cognitive discovery group. No such relationship was found in the academic programmes. As Bissell points out, this finding fits closely with the different philosophies of learning involved – one viewing development as a set of 'inseparable components' involving social, emotional and intellectual skills; the other breaking development into a series of discrete and specific skills that can be taught as a set of precise behavioural objectives.

More recent studies by the Stanford Research Institute on the Follow Through programme have extended these findings (Stallings 1974, 1975). Here seven different approaches were examined for their effects at first and third grade. Preliminary results suggest that where programmes followed systematic instruction backed by positive reinforcement schedules, the result was higher scores on measures such as reading and mathematics; children also showed greater 'task persistence' – engaging in more sustained self-instruction. Where instruction was on a more open and flexible basis, children made greater progress on a non-verbal problem-solving test, had lower absentee rates, and showed greater independence from adults, and more cooperation with other children during their work. Interestingly, other results indicated that children in the more flexible classrooms tended to take responsibility for their success but not their failure, while the reverse was true in the more structured programmes, though the validity of attitude scales with this age of child must be open to doubt.

Though these findings are tentative, there is a growing pattern of results that fits closely with differences in programme objectives. Earlier studies may have obscured this feature by using overall measures of change in their search for the most effective form of preschool curriculum. These results indicate real dilemmas of choice; which skills do we really wish to boost, and which will prove most effective in the long run? The results suggest that short-run success could be counter-productive at a later stage.

## Evidence from the West Riding EPA
The West Riding Educational Priority Area Project was one of five action-research schemes supported by a three year grant (1969–71) from the SSRC and DES (Halsey 1972). A further two year grant (1972–4) was given to the West Riding to study the follow-up effects of preschool education. Unlike the other projects which were all at the centre of large cities, the West Riding area

was a small South Yorkshire mining town, with a stable and close-knit community. With the exception of a few professionals, the population was solidly working class, the majority employed in the local pits. Local schools reflected the stable community; staff turnover was exceptionally low, and education and teaching had considerable local prestige and support – though all schools had below average levels of performance.

The EPA programme was not only concerned with preschool; the West Riding project set up schemes at both primary and secondary level, including a multipurpose educational centre. There was also supporting research which examined levels of pupil performance and conditions in local schools. However, early education was the major activity. At the outset a review was made of the American experience (Little and Smith 1971), at that point entering its second phase with a decline in the early optimism, but the deepest pessimism yet to come. How far do the results from the limited series of experiments in the West Riding parallel the subsequent direction of American research? We review some of the results, taking the issues of curriculum content, progress in the follow-up period, and the effects of sustained intervention in turn. In comparison to many American schemes, this was a micro-level study, involving one area, a few schools and less than 200 children – a further reason for trying to relate the results to larger studies.

First it is necessary to give a more detailed account of the experiments. Like many American studies the initial assumption was that certain forms of preschool intervention could accelerate early learning, and bring average levels of performance closer to national norms. The immediate problem was seen to lie in the discontinuity between early socialization at home and the requirements of school, particularly marked in EPAs. The need was for a preschool curriculum that concentrated more on cognitive skills, and for this reason the traditional approach with its emphasis on social skills and free play methods was felt not to be adequate. However, the poor follow-up results from highly structured programmes seemed to suggest that these were not the answer on their own. The conclusion stressed several ingredients; first, addition to the traditional form of preschool curriculum, involving more highly structured sessions aimed principally at language development. The project experimented separately with the individual tutoring sessions developed by Marion Blank

(1973), and, in conjunction with other EPA projects, with the Peabody Language Development Kit (PLDK) which provides for short language 'lessons' to be inserted into the normal curriculum. Second, parental involvement was emphasized, as a way of strengthening the effects of preschool. And, third, some form of follow-up programme in school was seen to be necessary if gains were to be maintained. Preschool intervention on its own was not sufficient, and did not rule out the need for action on other fronts, in schools, in adult education, in social and community work or in broadening occupational choice.

The West Riding project area was totally without any form of preschool provision; this allowed a series of experiments to be developed from scratch. In each case programmes were set up by defining catchment areas and age groups to take part, beginning with those areas where the schools showed the lowest levels of performance. Individual families were approached and asked if they would like to participate. The intention was to achieve universal cover for two reasons; first to avoid any selection on individual criteria, and second to ensure as far as possible that the complete age cohort entering reception class in a particular school had experienced preschool education.

Table 15.1 sets out the groups established during the three year project in three separate 'waves', showing the programme followed, the tests used and the extent of follow-up. With the exception of the 'pilot' phase, all groups met on a half day basis throughout the school year. The pilot phase ran part-time for ten weeks during the summer, and a control group was only established at the 'post-test' stage to assess the impact of preschool on parental attitudes. Test scores were collected up to four years after the end of the experiment. In the second wave, children from two different school catchments were involved in two parallel experiments: in one case a comparison between normal nursery, and nursery with the addition of PLDK lessons; and in the other, a three way comparison between normal nursery, nursery plus the PLDK, and what we have called the 'hybrid' group – with strong educational content, parental participation and individual tutoring sessions. In both experiments children were randomly allocated to the groups before pretest scores were collected. Each school had a morning and afternoon group – one following normal nursery, the other normal nursery with the addition of the PLDK. The same teacher in each school was responsible for the

Table 15.1 *The groups involved in the West Riding preschool experiment, the tests administered at each testing point, and sample sizes*

| | Sample size | Summer 1969 | Autumn 1969 | 1970 | 1971 | 1972 | 1973 |
|---|---|---|---|---|---|---|---|
| **WAVE 1 Pilot scheme** | | | | | | | |
| Experimental group | 18 | EPVT pretest | EPVT posttest | EPVT at end of reception year | EPVT | EPVT | EPVT |
| Control group | 17 | * | EPVT posttest | EPVT posttest | | | |
| **WAVE 2** | | | | | | | |
| School A nursery group | 21 | | | | | | |
| School A PLDK group | 21 | — | EPVT, RDLS, pretest | EPVT, RDLS, posttest | EPVT, RDLS, end of reception year | Boehm | Boehm |
| School A hybrid group | 20 | | | | | | |
| School B nursery group | 26 | | | | | Boehm, Daniels & Diack | Boehm, Daniels & Diack |
| School B PLDK group | 23 | | | | | | |
| **WAVE 2 follow up** | | | | | | | |
| Language group | 20 | — | — | EPVT, RDLS, pretest | EPVT, RDLS, posttest | Daniels & Diack | Daniels & Diack |
| Control group | 20 | — | — | | | | |
| **WAVE 3** | | | | | | | |
| Language group | 12 | — | — | | | | |
| Control group | 13 | — | — | | | | |

EPVT = English Picture Vocabulary Test.  RDLS = Reynell Developmental Language Scales.

* The control group for the pilot scheme was selected from another school for a separate study not reported here, of parental attitudes to preschool education.

morning and afternoon group. The fifth group was run directly by the EPA project.

In the third wave an attempt was made to assess the individual tutoring programme separately from other elements in the 'hybrid' programme by dividing the group randomly in two. During the same school year as the third wave, some children who had participated in the second wave, then attending infant school, were involved in an additional year of intervention with individual language periods during school time; most of these children therefore received two years' continuous extra provision of one type or another.

Evaluation was based in a limited set of tests, though these focused on language development. The English Picture Vocabulary test (EPVT) was used on every occasion, and this provides the complete longitudinal series. The Reynell Developmental Language Scales (RDLS), yielding two standardized subscores, 'comprehension' and 'expressive language', are more applicable than EPVT in assessing gains over the period of intervention, but have shorter age span with a ceiling of six years. The Boehm test of basic concepts and the Daniels and Diack Standard Reading Test 1 were introduced at the end of the follow-up period as tests of skills relevant at infant and junior school levels, but they do not contribute to the longitudinal series of results and are not reported here.

## Curriculum content

The pilot phase gave an opportunity to test out some of the project's preschool ideas, such as parental involvement and individual tutoring. The intention in the second 'wave' – the main experiment – was to compare different forms of preschool curriculum, not to assess the advantages of attending preschool against staying at home. There was thus no 'non-treatment' control. The assumption was that preschooling was desirable for many reasons but one form would prove more effective than another.

In fact the results over the nursery year on the RDLS and EPVT showed no such decisive differences between one group and another (Figures 15.1, 15.2 and 15.3). The picture is slightly clouded by unexpected differences in initial group scores, despite random allocation. However, on the EPVT and Reynell comprehension scale, there are no significant differences between groups as a result of the programme; in fact the two groups from School B remain rigidly in step on all those measures, this time a tribute to the original random allocation, though not to the effectiveness of the PLDK programme. All programmers were equally effective judged against test norms – or equally ineffective if this improvement is attributed to a test artefact. The only significant differences found were among the three groups in school A on the expressive language scale. Here the two experimental programmes made higher rates of gain than the normal nursery group, which in fact dropped back in standardized terms as it did also on the EPVT over the same period. But note that this lapse was made good the following year, and there is once again no difference between the three groups.

Perhaps on the expressive scale we are seeing the effects of the special programmes, which may have encouraged children to be more responsive to adults' questions, in line with the formal question and answer situation that both the PLDK and the individual sessions created. However, the PLDK did not register even this impact in School B; here one explanation seems to be the

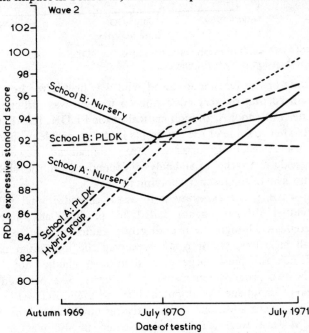

*Figure 15.1* Graph of mean standard scores on RDLS expressive language subtest for wave 2.

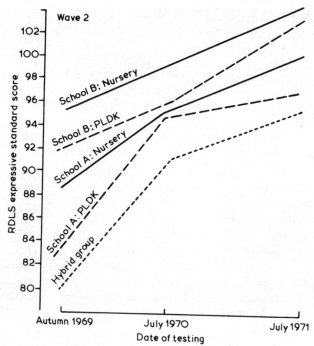

*Figure 15.2* Graph of mean standard scores on RDLS verbal comprehension subtest for wave 2.

attitude of the teacher involved who was hostile to the kit, and convinced that nursery education on its own was sufficient. The teacher in School A, though critical of the PLDK, felt that it had alerted her to some of the children's language problems, and given her useful teaching materials. Any programme, however well designed, is in the final analysis dependent on the skills and enthusiasm of the teacher applying it.

In wave 3, the experiment to assess the additional effects of individual tutoring again failed to produce any significant differences. Though the overall group gained against test norms on all measures, children who received the individual tutoring made no significantly larger gains than those who had not.

The two programmes tested in this series of experiments, the Peabody Language Development Kit and the individual language sessions, were additions to the existing curriculum, and in one case at least were hardly implemented in the way intended. Though larger-scale programmes may have made more impact, what is striking about the present results is the way most groups, including the normal nursery, tended to gain against test norms – the main exception being on the expressive language scales. Significant results favouring the special programmes were restricted to specific measures – a finding that perhaps echoes some of the American results.

*Follow-up results*

The experiment in wave 2 was originally intended to assess the comparative effects of different preschool programmes. By the end of the reception year, however, as Figure 15.3 shows, any differences observed during the nursery year had disappeared; the groups cluster closely by school. Tests were administered each summer for a further two years, providing a series of scores on the EPVT (preschool version and Level 1) from the start of the nursery year to the end of infant school. Do these provide any evidence of follow-up effects from the preschool experience? Two admittedly weak points of comparison are possible – with test norms – or with the results from the baseline testing programme carried out in 1969.

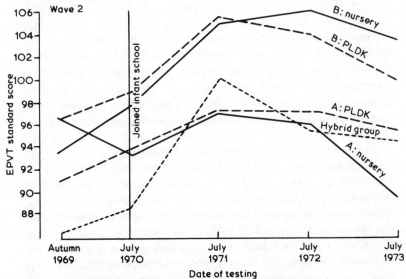

*Figure 15.3* Graph of mean standard scores on EPVT for wave 2.

Here EPVT scores were obtained for all children in the local primary schools, giving a profile of scores for different age cohorts. Given the stability of the area and the fact that there were no preschool facilities in the area before 1969, these baseline scores provide some form of benchmark to assess the progress of children in wave 2 who represented a complete age cohort from the catchment areas of two schools. It should be underlined that this is to make a comparison between children of different ages tested once only and one group of children tested at regular intervals. Any superiority by the second group could thus be the result of test familiarity, or there could be several other explanations.

The pattern of scores traced out by children in the pilot phase throws some light on this possibility. Though showing an initial gain during the short summer preschool and reception year, the pilot group subsequently declined, and the fourth year in school were down to the 'baseline' level, though they had by then taken the EPVT six times in just over four years. Test familiarity does not seem in this case to have boosted scores in the long run.

Children in wave 2 follow the same score profile as the baseline results. Though generally gaining in standardized terms during the nursery years, they show a further rise during the reception year – a phenomenon found in EPVT scores in all the EPA areas studied (Payne 1974), though one which has not yet been satisfactorily explained. Over the next two years, the groups gradually decline against test norms, but the same pattern is shown even more sharply in the baseline scores. Thus at the end of the third year in school the nursery groups remain above the corresponding mean baseline score for their age group. The difference varies by school; groups from school A were only marginally higher than the equivalent baseline point for school A, but the two groups from school B were some eight points higher, the difference being statistically highly significant.

These are hardly conclusive findings on the long-term effects of preschool education in view of the limited scope of the test involved, possible changes in the area or its schools between 1969 and 1973 and the series of intervening factors between nursery and the end of infant school. However, the pattern of results in the follow-up scores for school B suggests that there is a trend to be explained. Perhaps this school, where the experimental nursery year produced only small changes in mean score, was able to build on this development more effectively during the early school years. Subsequent events of this kind make it virtually impossible in small studies to disentangle the effects of earlier and later experience. Indeed, the very attempt to do this may be misplaced, as it seems to be based on a once-and-for-all mentality which assumes that early results will shine through.

In an attempt to test the relative effects of preschool experience and school differences, a regression analysis was run on the scores of the PLDK and nursery groups to estimate the variance in test score contributed successively by school, programme and the interaction between them. [3] The criterion used was the increment in EPVT standard score between pretest and final testing. Because of the non-orthogonality of the design, multiple regression rather than analysis of variance was applied. Of the gain score, the amount of variance accounted for by school, programme and the interaction between them was 21 per cent; school contributed 7 per cent, programme a further 1 per cent; the interaction between them added another 13 per cent, an amount which achieves a high level of significance. In other words, differences between schools appear to be related to differences in the way the programmes affect later performance. 'School differences', it should be pointed out, include any systematic differences between intakes as well as characteristic features of the schools.

*Follow-up intervention*
From the start, the West Riding project had accepted that a single year of preschool intervention was unlikely to have long-term effects. Ideally a sustained programme was needed. Following the main preschool experiment in wave 2, a chance arose to mount a further programme in the reception year, when one of the infant schools involved (school A) agreed to try out the individual language scheme with some of its pupils. Most of these had already experienced a year of nursery education, some in the PLDK or hybrid group with its individual programme. Using the end-of-nursery score as a pretest measure, it was possible to construct a further experiment by creating two groups randomly allocated from pairs matched on the EPVT. One group had regular individual sessions with members of the project team, mature teacher training students and occasionally teachers in the school, throughout reception year. The other group followed normal class work.

Figure 15.4 presents the EPVT results for the experimental and two subsequent years. During the first year the experimental group had made significant gains on the EPVT and Reynell comprehension scales, and larger, though not significant, gains on the expressive scale. One year after the programme had ended, however, the differences on the EPVT disappeared; and after two years the experimental and control groups again had almost identical scores. These results suggest that an additional year of intervention could make an additional impact, though as with preschool further intervention seems in no way to guarantee a long-term gain in performance.

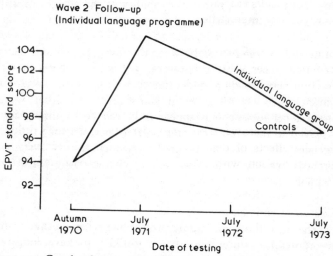

*Figure 15.4* Graph of mean standard scores on EPVT for the wave 2 follow-up individual language programme.

In comparison to many American studies, the West Riding experiments are at micro-level. Detailed exposition of the results rapidly leads to explanations of the position in a particular school, or the attitudes of a particular teacher. Though these may explain events in the West Riding, they are of less use in generalizing to other areas, unless they also reflect more general arguments, such as the importance of teachers' attitudes and enthusiasm, or 'staff model', as Weikart terms it. Results from the West Riding suggest that a search for the universally 'most effective' programme is misplaced effort. Different programmes sometimes have similar effects, and at others may promote particular skills, depending on the way they are implemented. There is tentative evidence of sustained impact from early education, though this may be more a product of subsequent schooling than of the original stimulus. A further year's intervention produced additional gains, though these did not last more than a year after intervention had ceased.

The West Riding experiments suffered several of the same weaknesses as the early phases of American programmes; they were based on optimistic assumptions about the ease with which gains could be achieved and maintained by early intervention, and partly for this reason they used a limited number of measures to assess effects. Despite these weaknesses, the findings give some hint of the results that could be achieved by carefully designed and evaluated early intervention and follow-up studies.

## Deprivation and preschool intervention

Lying behind the shifting debate on the effectiveness of preschool education has been a changing view of deprivation and disadvantage. Much has been made of the 'shot in the arm' philosophy of early programmes, where the right form of experience during the preschool years was thought to guarantee a lifetime's protection from the effects of deprivation. This approach virtually identified deprivation with inadequate early socialization. Though this early socialization may itself be seen as the product of other social forces, it is the mediating element that determines the child's later development. To change the metaphor, the 'deprived child' is seen to have collapsed on the sidewalk, a helping hand at the right moment to set him on his feet will ensure he continues on his way unaided.

Experience has undermined this simple model. The child collapses again once the helping hand is withdrawn. What emerges is a far more complex picture, where the child and his family are continuously subject to a series of pressures, in housing and employment, environment or social relationships. These have their effect on patterns of early socialization, but they continue to influence growth and development either through the family, or through school, peer group and neighbourhood experience. The disadvantaged are those who have systematically lost out in the struggle for scarce resources; different patterns of socialization and different emphases in early development are in part a necessary adaptation to these harsh realities, rather than purely the result of 'cultural differences'.

Preschool intervention alone makes no impact on these factors;

not surprisingly, once it is withdrawn they reassert themselves. Clearly, preschool effects can be extended if preschool covers greater areas of social life, either by moving as close as possible to some form of total care, or by seeking to influence the behaviour of parents as well as children. 'Ecological intervention' is the logical next step, followed by a sequential series of programmes from pre-parenthood through the early years to end of the primary stage – a formidable and costly series of projects.

The danger in following this line of argument is that we will move from a position where there seemed to be a single key, to one where it appears we have to change everything to have any effect at all. Clearly some factors must be more important than others. And here the changing view of deprivation has pushed education from the centre of the stage. If we want changes in income distribution, housing or job opportunities, then it is best to work to change them directly, rather than hope for indirect effects through education. Yet education still retains two important roles. First, it is a means by which disadvantaged groups can become aware of inequalities, and more likely to press effectively for change; and second, even if opportunities are made more equal, groups who have missed out in the past will require the skills to take advantage of and maintain these gains. We still require an educational strategy, though one that sets education firmly in its social context.

The strategy, we would suggest, will be defined on two main dimensions. First, which are the key areas for educational intervention? The preschool years, the main transitional points in schooling, and certain forms of adult education appear likely candidates for concentrated resources. Second is the question of social context. Our work in a stable mining community in South Yorkshire convinced us that the successful results could in part be attributed to this stability in school and community. The experience of trying to operate a home-visiting programme during a national pit strike showed how quickly the position could change. Under these conditions the immediate concern of families was survival, rather than discussion about the most suitable books and toys for young children. In many deprived parts of cities these crisis conditions are frequent; the need is for immediate help, not the long-term investment promised by early education. Analysis of the particular local context must determine strategy.

However, under stable conditions, the evidence we have

reviewed suggests that preschool intervention can make an impact, and with the right support this can be maintained for considerable periods. The psychological and developmental reasons for preschool intervention were oversold, particularly the simple view of deprivation they assumed. But there remain strong social and educational reasons for arguing that the preschool years present the best opportunity to raise levels of performance. For a start, it is a chance to intervene before any cycle of failure has built up, when there is a high degree of parental interest and enthusiasm. However, we have to treat preschool intervention as a stage; advance here does not guarantee long-term success. It is only a stage; but one that provides many of the preconditions for successfully negotiating the next phase. Later experience can either support and build on this progress, or it can undermine it.

The early optimism about the effects of preschooling was matched by an equal optimism about the ease of translating research findings into practice. The belief was that research and experiment would conclusively demonstrate which were the best options, leaving the policy makers only the simple task of wider implementation (Cohen and Garet 1975). The result has been almost the opposite. Far from the policy question being answered, it has been made infinitely more complex; new forms of preschool intervention have been developed, different approaches shown to have different but equally valid results, and effectiveness to be as much a product of local conditions as of any experimental programme. Policy makers naturally want clear answers from research that can be applied across the board. The results are the sudden swings in fashion as one overall conclusion is overthrown by another – usually its opposite. The problem is to win acceptance that policy research is likely to open new questions and raise further dilemmas, placing the choice of policy options squarely back where it belongs. Effective policy research, we suspect, should make policy making more difficult, not less.

**Notes**

1 For example, Shirley Williams in a speech at Ditchley Park, 1969, recounted how as Minister at the DES she had called for the findings of preschool research, only to be told that there was little or nothing available.

2 The claim is certainly a curious one. What does it mean in quantitative terms to say that a four-year-old has 'half the intelligence' of an adult?

3 We are grateful to Barbara Tizard for suggesting the analysis and to Angela Skrimshire for carrying it out.

# 16 Genetics and intelligence: The race argument

## W. Bodmer

(From *Times Literary Supplement*, August 1973.)

Inheritance and the workings of the mind have no doubt fascinated and intrigued mankind since time immemorial. Allegiance to one's own group – be it family, village, country, race or even specie – seems to be almost instinctive and with this instinct comes its complement, racialism. Put these three ingredients – genetics, intelligence and race – together, and you have one of the most emotionally charged topics of the day linking science and social affairs. The scientific background needed for a proper discussion of the extent to which genetic factors can contribute to racial differences in intelligence is complex and highly technical. Folklore on inheritance of behaviour and preconceived notions on racial differences are, however, so widespread as to make difficult technical discussions a poor answer to politically motivated arguments. The borderline between science and politics may be hard to define, but the scientist's responsibility to the public is most important and the need to explain the scientific issues as simply and as objectively as possible is paramount.

The possibility that racial differences in ability are inherited has been discussed sporadically for many years. Though writers before Galton and Mendel, such as the nineteenth century French diplomat Conte de Gobineau, who was a great proponent of the superiority of 'the German race', could hardly have been writing on a firm scientific footing, some of the early human geneticists of this century were hardly less ardent in their claims. The most recent eruption of this issue is, of course, due to Jensen, a Professor of Educational Psychology at the University of California in Berkeley, closely followed by Shockley, the Nobel prize winning physicist and co-inventor of the transistor in the United States, and the psychologist Eysenck in this country. The hub of recent controversy is an article by Jensen called 'How much can we boost IQ and scholastic achievement?' published in 1969 in the *Harvard Educational Review*. This article is the centrepiece of Jensen's book *Genetics and Education*, published in autumn 1972, while his more recent book *Educability and Group Differences* is essentially a very much expanded version of the *Harvard Educational Review* article.

The essence of Jensen's argument is as follows. He starts from the assumption that IQ as determined from a variety of more or less standard tests is a quantitative measure of some aspect of human intellectual behaviour. Genetic studies within the white population of the United States and of Europe, and in a few cases also within the United States black population, suggest that a major part of the variation in IQ, perhaps as much as 80 per cent, may be genetically determined. Some racial groups, notably US whites and blacks, differ very substantially in their distribution of IQ. The mean IQ of US whites, for example, is in many studies ten to fifteen points higher than that of US blacks. This is comparable to the standard deviation of the IQ distribution, which is fifteen points, and which means that about 95 per cent of the typical US white population, whose mean IQ is 100, have IQs between 70 and 130. Jensen's argument then continues with an assessment of possible environmental factors which can influence IQ, the conclusion being that none are known which could explain such a large racial difference. Thus, he would say, since IQ is largely genetically determined within whites and blacks, there is most likely to be a substantial genetic component for the race difference. The sociopolitical overtones to the argument are based on the idea that IQ, which does vary appreciably with socioeconomic status as usually defined by sociologists and economists, is a major determinant of success in our society. So it is important to know how much of the race-IQ differences really are genetic.

Stated baldly in this way, Jensen's not-so-novel thesis may sound simple and logical, but is it really so? Embedded in the argument are many scientific concepts and questions whose analysis does not flow so easily. What is IQ and what does it really measure? How can one sort out the inherited and environmental components of a complex, quantitative character such as IQ? What is a race and how can it be objectively defined? Can one extrapolate from knowledge of genetic components within a race to genetic contributions to the mean difference between two races? Is it really true that IQ is a major determinant of success in our society? And last, but by no means least, even if the genetic component of the race-IQ differene were established, what practical use could be made of this information?

Many scientific papers and books have been written on these questions to which the interested reader must turn if he is really to judge Jensen's arguments for himself. Each of the two books by Jensen concludes with a bibliography that covers much of the subject, including Jensen's own writings. Though the bibliographies are not annotated, Jensen's comments on them throughout the books leave one in no doubt as to which he would recommend and so the reader can choose among them according to his tastes.

Intelligence must not be confused with IQ as measured by an IQ test. The psychologists have, over the years, devised many types of tests, various combinations of which are used in any given situation to assign an IQ to an individual. This one number can hardly be considered the complete definition of intelligence, though it presumably measures some component of intellectual ability. Psychologists seem to be divided into two schools – namely, those like Jensen and Eysenck who believe that there is some basic property of the intellect called $g$ (for general intelligence) that is measured by an IQ test, and those who believe that intelligence is multidimensional and cannot properly be measured by a single number. The geneticist can stand aside from these arguments and simply use the number given him by the IQ tester for his genetic analysis as he would for less contentious measurements such as height or weight. Thus, even though the real significance of the IQ measurement may be in doubt, the geneticist can still try and answer questions about the extent to which whatever it is that is measured by an IQ test has a genetic component.

A major key to Mendel's success in uncovering the basic laws of inheritance was to work with simple, easily defined characters. The subsequent development of genetics and molecular biology, leading to the chemical definition of the gene in terms of the DNA molecule and of its mode of action through the specification of proteins, has similarly depended on working with precise well-defined biochemical differences. Given a simple character that is determined by one or at most a small number of identified genes, its patterns of inheritance can be unequivocally defined. Many genes are known which cause severe mental retardation and so, of course, have a catastrophic effect on IQ. These genes are, however, quite rare and so have no substantial impact on the overall distribution of IQ in a population. IQ, as measured by an IQ test, must be a composite and complex character. Its expression is dependent on a combination of the effects of environmental factors and the product of many different genes, each of which has on average a very small effect on IQ and so cannot be individually recognized. As a result, one has to resort to complex statistical analyses to sort out the relative contributions of heredity and environment to IQ. The aim of these analyses is to assess the extent to which relatives tend to be more like each other than they are to unrelated or to more distantly related people. The results are usually expressed in terms of a quantity called the *heritability*, which, loosely speaking, measures the proportion of the variation in IQ that can be ascribed to genetic factors. If the heritability were zero then all variation would be environmentally determined, while if it were one all variation would be genetic. The true answer nearly always lies somewhere between these two extremes. Heritability is not like a physical constant. It is just a very rough and ready statistical measure of a property of a quantitative character, such as IQ, in a given population. As the population and its environment changes, so does the heritability. Methods for measuring heritability were originally devised for plant and animal breeders, who can do controlled crosses at will and can, to a fair extent, control the environment in which their crosses are made. The human geneticist is in a much more difficult position as he can do neither of these things. Even worse, the most convenient types of relatives to study – namely, parents and offspring and brothers and sisters – are, for the most part, found in the same home environment. This means that one can never really be sure whether similarities between parents and offspring or between brothers and sisters are due to the genes they share or to the common environment in which they live. The only partly satisfactory answer to this problem is to study adopted children (especially identical twins separated at or near birth) who do not share a common environment with their immediate relatives. Most studies on IQ, the majority in white American and European populations, suggest a substantial heritability for IQ, with estimates ranging from 40 to 80 per cent, according to the sources of the data and the types of analysis performed. This certainly suggests that there are substantial genetic differences in whatever it is that is measured by an IQ test.

Apart from identical twins, all people look and behave differently. The outward physical features by which we distinguish people are paralleled by simple inherited differences such as the blood types, which can be defined at a chemical level. One of the most striking results of population genetic studies over the last ten years is the demonstration of just how much genetic variation there is in human (and other) populations Even the genetic differences one can test for now, which can only be a minute fraction of all those that exist, are enough to make the chances of finding two unrelated identical people much less than one in a million. It will not be long before it should be possible to give everyone a unique genetic typing in terms of well-defined, simply inherited chemical characteristics. The potential for genetic differences between individuals is staggering, even within a family. The numbers of genetically different types of sperm and egg which any one individual could in principle produce is many millionfold more than the numbers of humans that have ever lived. This extraordinary genetic uniqueness of the individual must apply to all his attributes, including intelligence, and is presumably the basis for the genetic component to IQ.

The existence of genetic variability in a character does not in any way imply the absence of environmental effects. Some of the most severe simply determined genetic diseases, such as phenylketonuria, can be almost completely reversed by an appropriate, though complex, diet. There is, as might be expected, plenty of evidence for environmental factors affecting IQ. Amongst the most striking are the consistent approximately five point difference between the IQs of twins and single births, and the extraordinary IQ differences between adopted children and their biological parents that have been reported in some adoption studies.

Most IQ studies on different racial groups have used a sociological definition of race. In the USA, for example, children of mixed black–white marriages would be classified as black, though even the black parent could be from a mixed marriage. Genetic studies, in fact, show that all US blacks have, on average, between 10 and 30 per cent white ancestry. To a biologist a race is just a group of individuals or population which forms a recognizable subdivision of the species. The group is identified by the fact that individuals within it share characteristics which distinguish them, at least to some extent, from other subgroups. These characteristic differences are maintained because individuals belonging to a group are most likely to find their mates within the group. Traditionally, human races were defined using outwardly obvious features, such as skin colour, hair colour and texture, facial and other physical characteristics, whose inheritance is still poorly understood because of their relative complexity. Nowadays, geneticists agree that the only objective biologically valid approach to defining races is in terms of simply inherited well-defined differences, such as the blood types. The frequencies of the various genetic types often differ from one population to another and it is these frequencies which are used to characterize population groups. The group is thus a statistical concept and the borderline between races may therefore be blurred and hard to define. Any particular genetic combination may be found in almost any race, but the frequency with which it is found will vary from one race to another. Some features, of course, such as skin colour, show very marked differences between races, which are most probably adaptations to different climatic conditions and resulting life styles. But these are the exceptions rather than the rules. Studies of genetic differences within and between races show that at least 60 to 70 per cent of the overall genetic variation in man occurs *within* races.

Since the race is a statistical concept, the question of a genetic component to a race difference is logically quite different from that of genetic differences between individuals. The difference between two races is a difference between *averages*, and reflects differences in the frequencies of genetic types in the two races. Individual genes vary in frequency between races for many reasons, some of them just due to chance, and some due to the action of natural selection in one form or another. Jensen and others have argued that because such differences in the frequencies of individual genes exist between races, why not also genetic differences in IQ? But IQ, as we have seen, must be determined by many genes. If there is to be an average genetic difference in IQ between two races, then there must be a tendency for those genes which on average increase IQ to accumulate in one of the races, and/or a similar tendency for those genes which decrease IQ to accumulate in the other race. This is only likely to happen simultaneously for a large number of genes if IQ is in some way itself subject to the action of natural selection. Nobody, however, has to my knowledge offered a convincing model for the way in which IQ test measurements, as we obtain them nowadays, could have been connected with increased ability to survive or to reproduce during the evolutionary divergence of the present-day races of man.

A number of geneticists have emphasized the fact that there is no logical connection between genetic components determined within a race and the extent to which a difference between races has a genetic component. The environmental factors which distinguish two racial groups, such as the blacks and whites, may be quite different from the environmental variations found within either race. Cultural differences between races may simply have no parallel within races. While Jensen freely acknowledges that within-population heritability may have nothing to do with genetic differences between populations, once having acknowledged this point, he seems to feel that he is then entitled to ignore it. Nowhere in either of his books does he discuss this absolutely crucial point with any conviction.

Jensen's two books are, apart from the prefaces, not easy reading even for the relatively initiated. There is much statistical methodology, which is not helped by a rather heavy style in which sentences such as the following are not uncommon:

> Because variance in achievement test scores reflects a larger gain component at any given time than do intelligence tests, which are designed to reflect the consolidation factor, one should expect populations that differ on the average on intelligence measures to differ significantly less on achievement measures at any cross-section in time, and this has been found to be the case.

Jensen says in his preface to *Educability and Group Differences* that 'it was written with mainly behavioural scientists and educational researchers in mind'. I hope he is right in assuming that they have the appropriate background to deal with the material. Certainly, these are not books for the layman.

In *Genetics and Education* only the preface is new. In addition to the *Harvard Educational Review* article, there is a long paper on 'A theory of primary and secondary familial mental retardation'; three more or less technical papers on estimation of heritability, one of which I am sure even Jensen would admit was obviously written before he had become fully conversant with the techniques of quantitative genetics; and a short ethical note on the right to do research on whatever one wants

Jensen's theory of mental retardation is that there are two hierarchical components to mental retardation, one connected with 'associative ability' and the other with 'cognitive ability'. They are hierarchical in the sense that the first is required for the second, which latter is most closely related to IQ test performance. The two levels of ability are assumed to be under independent genetic control. While the theory may well have some heuristic value, it seems to me, as a biologist and geneticist, a rather simpleminded dichotomy of a complex problem.

The preface to *Genetics and Education* recounts the history of his *Harvard Educational Review* article and the ensuing problems, criticisms and castigations Jensen has faced. Having told how he gave a copy of his article to a staff writer from the *US News and World Report*, Jensen protests that he was surprised that his conclusions raised such a public furore. He should not have been surprised. After all, few scientists give advance copies of their articles to journalists! Communication of science to the public is an important and difficult task. The non-scientist is generally not trained to deal with scientific arguments based on hypothesis and counter-hypothesis. A good reasoned and well-qualified scientific statement does not make good copy for newspaper headlines.

Jensen rightly complains of the treatment which he and others, such as Eysenck in this country, have received at the hands of those who, while protesting they champion freedom, do not seem to support freedom of speech. It is surely one of the great strengths of our democratic society that it can tolerate open discussion of all shades of opinion – scientific, political or otherwise. Those who seek to silence Jensen, Eysenck or Shockley do not help the fight against racial prejudice which they claim to be supporting. Having said this, however, I am not tempted to compare Jensen with Galileo or any other of the more famous scientific martyrs. Jensen's scientific contributions hardly seem to match those of these more illustrious scientific forebears. Nor have his writings been suppressed. The *New York Times Magazine* devoted practically a whole issue to his ideas and his Harvard Education Review article was read into the US Congressional Record in its entirety.

*Educability and Group Differences* is entirely devoted to the problem of the inheritance of IQ and the genetic component to the race-IQ difference. It is an extensive and well-documented review of the field. In addition, however, to its complexity, the style has a fighting defensive flavour to it, almost every chapter opening with a salvo against the environmentalist. Just to set the scene, the second chapter is entitled 'Current technical misconceptions and obfuscations', while in the preface we are admonished to *'think genetically'*, as if to imply that all right minded geneticists are bound to come to the same conclusions as does Jensen: 'The problem on both sides is fundamentally a matter of ignorance, the cure for which is a proper education about genetics.'

The book takes us painstakingly through definitions of heritability, statistical approaches to the analysis of data and all the various factors which might influence IQ. There is a touching faith in complex statistical procedures which Jensen seems to believe give to IQ almost the same precision as a measurement made with a ruler or balance. Much emphasis is, of course, placed on the estimation of the heritability of IQ and on its high value, and also on the apparent inadequacy of environmental explanations for the race-IQ difference. Jensen somehow appears to be obsessed with the need to find single factors in the environment which account for a substantial fraction of the fifteen point IQ difference. Repeatedly we are told that some given effect, be it schooling, nutrition, socioeconomic status or any one of at least a dozen different factors, can explain at most a very small fraction of the difference, perhaps only one or two IQ points. The precision of the data, in any case, hardly excludes effects of this order of magnitude. Surely he must appreciate that such effects are cumulative and may combine to explain the whole of the difference, even though individually any one of them has only a very small effect. Jensen would, I suppose, argue that the effects he mentions are not independent, but he does not properly document the evidence for this. Others at least as well versed in these questions, such as Jencks, simply do not agree with Jensen on these issues.

A few years ago my colleague Luca Cavalli-Sforza and I, in assessing the evidence on genetic and environmental influences on race-IQ differences, came to the not very startling conclusion that current techniques and data could not resolve the question. While not excluding the possibility of a genetic component, it seemed to us that the IQ differences could be explained by environmental factors, many of which we still knew nothing about. I do not believe the situation has changed since then. Jensen, however, seems to have changed his tune somewhat. In his *Harvard Educational Review* article he wrote:

> So all we are left with are various lines of evidence, no one of which is definitive alone, but which, reviewed all together, make it a not unreasonable hypothesis that genetic factors are strongly implicated in the average Negro–white intelligence difference. The preponderance of the evidence is, in my opinion, less consistent with a strictly environmental hypothesis than with a genetic hypothesis, which of course does not exclude the influence of environment or its interaction with genetic factors.

The first paragraph of his summary of the final chapter of *Educability and Group Differences* ends with the following statement: 'All the major facts would seem to be comprehended quite well by the hypothesis that something between American Negroes and whites is attributable to genetic factors, and the remainder to environmental factors and their interaction with genetic differences.' Jensen now seems sufficiently convinced of his hypothesis to quote a definite figure for how much of the race-IQ difference is genetic.

Is IQ really so significant as a determinant of success in our society? What action flows from the knowledge of the existence of a genetic component to the race-IQ difference? Nowhere in Jensen's two books are these two all-important questions properly answered.

The answer to the first question given by Jencks in his very readable book *Inequality* (1972) is a resounding 'No'. This book, incidentally, is a welcome complement, for some no doubt an antidote, to Jensen's writings. Many factors are involved in the determination of economic and social success in our society, and measured IQ is only a small part of the story. This does not deny the existence of very significant IQ differences between different socioeconomic groups, but only emphasizes that the genetic component to these differences must not be exaggerated.

As to the importance of the race-IQ difference, apart from policies of selective breeding, which I am sure Jensen would not support, there are such questions as the relatively high number of black as compared to white children in schools for the educationally subnormal. Quotas based on the racial distribution in the population might prevent black children from receiving special educational programmes when they were really needed, and this is to be deplored. But the remedy has nothing to do with *genetic* differences between the races. The problem is the same whatever the cause of the differences. In a society which professes to be free of racial prejudice people are treated as individuals not as members of groups, however defined – whether by race, religion, sex or otherwise. The only practical case I can see for studying the genetic component of the race-IQ difference is to answer the question that has been raised, and this is a weak case indeed.

There is much to be learned about human behaviour and the workings of the mind and I personally am confident that much benefit may come from such further knowledge. I am not convinced, however, that there are absolutely no limits to basic research, though I do believe that, on the whole, it must be allowed a free rein. We do, after all, subscribe to ethical codes for medical and behavioural research on human subjects and this clearly sets some limits to what is allowable. Planned support of major applied research endeavours is quite another matter, and here I do not see the case espoused by Shockley, and to a lesser extent by Jensen, for crash programmes to answer such questions as the extent of the genetic component to the race-IQ difference. I believe that Jensen is an able scientist and an effective research worker. For my part, I just wish he would concentrate his talents in a more profitable area of research. Many interesting and important questions are raised by the existence of genetic variability for all aspects of human behaviour. Jensen himself makes the interesting point that, if individuals in a class learn at different speeds but all are taught at the same pace, then some will be bored, having finished a task early, while others may never have time to finish with one topic before moving on to the next. In this way, perhaps, the most basic skills will never be learnt by some, however much schooling they have had.

We do not at present know the answer to the question of how much is the genetic component to the race-IQ difference: we do not have the techniques at hand to find the answer and the answer does not seem to matter anyway. Let us then forget this question and move on to more important matters for our present day society.

# References

Abelson, P. H. (1969) Malnutrition, learning and behaviour. *Science,* **164,** 17.

Alland, A. (1971) Intelligence in black and white. In Brace, C. L., Gamble, G. R., and Bond, J. T. (eds) *Race and Intelligence.* Anthropological Studies No. 8. Washington D.C., Am. Anthrop. Assoc.

Allison, A. C. (1964) Polymorphism and natural selection in human populations. *Cold Spring Harbour Symp. Quant. Biol.,* **29,** 137–50.

Anastasi, A. (1956) Intelligence and family size. *Psychol. Bull.,* **53,** 187–209.

Anderson, C. A., Brown, J. C., and Bowman, J. J. (1952) Intelligence and occupational mobility. *J. Pol. Econ.,* **51,** 218–39.

Andrewartha, H. G., and Birch, C. L. (1954) *The Distribution and Abundance of Animals.* Chicago, Chicago UP.

Ardrey, R. (1966) *The Territorial Imperative.* New York, Atheneum.

Ardrey, R. (1976) *The Hunting Hypothesis.* London, Collins.

Atkinson, A. B. (1972) *Unequal Shares: The Distribution of Wealth in Britain.* London, Allen Lane.

Bajema, C. J. (1968) A note on the interrelations among intellectual ability, educational attainment and occupational achievement: a follow-up study of a male Kalamazoo public school population. *Soc. of Educ.,* **41,** 317–19.

Baker, J. R. (1974) *Race.* New York, Oxford UP.

Baldwin, A. L., and Baldwin, C. P. (1973) The study of mother-child interaction. *Am. Sci.,* **61,** 714–21.

Banks, J. A. (1954) *Prosperity and Parenthood.* London, Routledge & Kegan Paul.

Baraitser, M., and Evans, D. E. (1969) The effect of undernutrition on brain-rhythm development. *S. Afr. Med. J.,* **43,** 56–8.

Bateson, W. (1894) *Materials for the Study of Variation.* London, Macmillan.

Baughman, E. E., and Dahlstrom, W. G. (1968) *Negro and White Children: A Psychological Study in the Rural South.* New York, Acad. P.

Becker, G. S. (1960) An economic analysis of fertility. *Demographic and Economic Change in Developed Countries.* New Haven, Conn., Princeton UP.

Bereiter, C. (1969) The future of individual differences. *Harvard Educ. Rev.,* **39** (2), 310–18.

Bereiter, C. (1972) An academic preschool for disadvantaged children: conclusions from evaluation studies. In Stanley, J. C. (ed.) *Pre-school Programs for the Disadvantaged,* 1–21. Baltimore, Md., Johns Hopkins.

Bereiter, C., and Engelmann, S. (1966) *Teaching Disadvantaged Children in the Preschool.* Englewood Cliffs, N.J., Prentice-Hall.

Bettelheim, B. (1967) *The Empty Fortress: Infantile Autism and the Birth of the Self.* London, Collier-Macmillan.

Biesheuvel, S. (1952) The occupational abilities of Africans. *Optima,* **1**, 18.

Biesheuvel, S. (1972) An examination of Jensen's theory concerning educability, heritability and population differences. *Psychologia Africana,* **14** (2), 87–94.

Birdsell, J. B. (1958) On population structure in generalized hunting and collecting populations. *Evolution,* **12**, 189–205.

Bissell, J. S. (1973) Planned variation in Head Start and Follow Through. In Stanley, J. C. (ed.) *Compensatory Education for Children, Ages Two to Eight.* Baltimore, Md., Johns Hopkins.

Blank, M. (1973) *Teaching Learning in the Playschool.* Columbus, Merrill.

Blau, P., and Duncan, O. D. (1967) *The American Occupational Structure.* New York, Wiley.

Bloom, B. S. (1964) *Stability and Change in Human Characteristics.* New York, Wiley.

Blurton Jones, N. (ed.) (1972) *Ethological Studies of Child Behaviour.* Cambridge, Cambridge UP.

Bock, R. D., and Kolakowski, D. (1973) Further evidence of sex-linked major-gene influence on human spatial visualizing ability. *Am. J. Hum. Genet.,* **25**, 1–14.

Bodmer, W. F., and Cavalli-Sforza, L. L. (1970) Intelligence and race. *Sci. Am.,* **223** (4), 19–29.

Bosco, J. J. (1970) *Social Class and the Processing of Visual Information.* Final Report Project No. 9-/-041, Contract No. OEG-5-9-325041-0034(010). Washington, DC, US Dept of Health, Education & Welfare, Office of Educ.

Bourgeois-Pichat, J. (1965) Les facteurs de la fecondité non dirigée. *Population,* **20**, 383–424.

Bowlby, J. (1951) *Maternal Care and Mental Health.* Geneva, WHO; London, HMSO; New York, Columbia UP.

Bowlby, J. (1969) *Attachment and Loss,* vol. 1: *Attachment.* London, Hogarth.

Brace, C. L., Gamble, G. R., and Bond, J. T. (eds) *Race and Intelligence.* Anthrop. Studies No. 8. Washington, DC, Am. Anthrop. Assoc.

Breder, C. M., and Coates, C. W. (1932) A preliminary study of population stability and sex ratio of Lebistes. *Copeia,* 147–55.

Bressler, M. (1968) Sociology, biology and ideology. In Glass (1968), 178–209.

Bronfenbrenner, U. (1972) *Is Early Education Effective?* Cornell UP.

Brown, R. E. (1966) Organ weight in malnutrition with special reference to brain weight. *Develop. Medicine and Child Neurol.*, **8**, 512–22.

Bruner, J. S. (1975) Poverty and childhood. *Oxford Rev. of Educ.*, **1** (1) 31–50

Burnet, M. (1966) *Biology and the Appreciation of Life.* The Boyer Lectures, Australian Broadcasting Commission.

Burt, C. (1922) *Mental and Scholastic Tests.* London, King.

Burt, C. (1952) *Intelligence and Fertility.* London, Cassell.

Burt, C. (1959) Class difference in general intelligence. *Brit. J. Statist. Psychol.*, **12**, 15–34.

Burt, C. (1961) Intelligence and social mobility. *Brit. J. Statist. Psychol.*, **14**, 3–25.

Burt, C. (1966) The genetic determination of differences in intelligence: a study of monozygotic twins reared together and apart. *Brit. J. Psychol.*, **157**, 137–53.

Burt, C. (1971) Quantitative genetics in psychology. *Brit. J. Math. Statist. Psychol.*, **24**, 1–21.

Burt, C., and Howard, M. (1956) The multifactorial theory of inheritance and its application to intelligence. *Brit. J. Statist. Psychol.*, **9**, 95–131.

Cannon, W. B. (1932) *The Wisdom of the Body.* New York, North.

Carr-Saunders, A. (1922) *The Population Problem.* Oxford, Clarendon.

Carter, C. O. (1966) Differential fertility and intelligence. In Meade, J. E., and Parkes, A. S. (eds) *Genetic and Environmental Factors in Human Ability.* Edinburgh and London, Oliver & Boyd.

Cartwright, A. (1976) *How Many Children?* London, Routledge & Kegan Paul.

Cassen, R. H. (forthcoming) *Population, Economy, Society.*

Chance, M. R. A. (1967) Attention structure as the basis of primate rank orders. *Man*, **2**, 503–18.

Chase, H. P., Lindsley, W. F. B., and O'Brien, D. (1969) Undernutrition and cerebellar development. *Nature*, **221**, 554–5.

Cicirelli, V. G. *et al.* (1969) *The Impact of Head Start on Children's Cognitive and Affective Development.* Washington, DC, Westinghouse Learning Corp.

Coale, A. J. (1959) Increases in expectation of life and population growth. *Population Conference, Vienna*, 36–41. Vienna, International Union for the Scientific Study of Population.

Coale, A. J. (1969) The decline of fertility in Europe from the French Revolution to World War II. In Behrman, S. J., Corsa, L., and Freedman, R. (eds) *Fertility and Family Planning*, 3–24. Ann Arbor, U Michigan P.

Coale, A. J. (1973) The demographic transition. *International Population Conference*, **1**, 53–71. Liege, Internat. Union for the Sci. Study of Population.

Cochrane, S. H. (1975) Children as by-products, investment goods and consumer goods: a review of some micro-economic models of fertility. *Pop. Studies*, **29**, 373–90.

Cohen, D. K., and Garet, M. S. (1975) Reforming educational policy with applied social research. *Harvard Educ. Rev.*, **45** (1), 17–43.

Coleman, J. S. *et al.* (1966) *Equality of Educational Opportunity.* Washington, DC, US Dept of Health, Educ. & Welfare.

Conrad, H. S., and Jones, H. E. (1940) *Yearbook Nat. Soc. Stud. Educ.*, **39**, 97–141.

Conway, J. (1958) The inheritance of intelligence and its social implications. *Brit. J. Statist. Psychol.*, **11**, 171–90.

Cowley, J. J., and Griesel, R. D. (1965) The electro-encephalogram in low protein rats. *Psychologia Africana*, **11**, 14–19.

Cowley, J. J., and Griesel, R. D. (1966) The effect of rehabilitating first and second generation low protein rats on growth and behaviour. *An. Behav.*, **14**, 506–17.

Cravioto, J. and Robles, B. (1965) Evolution of adaptive and motor behaviour during rehabilitation from kwashiorkor. *Am. J. Orthopsychiat.*, **35**, 449–64.

Crow, J. F. (1969) Genetic theories and influences: comments on the value of diversity. In *Environment, Heredity and Intelligence.* Harvard. Educ. Rev. Reprint Series 2.

Crow, J. F., and Kimura, M. (1970) *An Introduction to Population Genetics Theory.* New York, Harper & Row.

Dahlberg, G. (1948) *Advances in Genetics,* II. New York, Acad. P.

Darlington, C. D. (1960) The future of man. *Heredity*, **15**, 441–2.

Darwin, C. R. (1871) *Descent of Man.* London, John Murray.

Darwin, C. R. (1872) *The Origin of Species.* London, John Murray, 6th edn.

Davis, K., and Blake, J. (1956) Social structure and fertility: an analytic framework. *Econ. Devel. & Cult. Change*, **4**, 211–35.

Davison, A. N., and Dobbing, J. (1965) Myelination as a vulnerable period in brain development. *Brit. Med. Bull.*, **22**, 40–4.

De Fries, J. C. (1964) Effects of prenatal maternal stress on behaviour in mice: a genotype-environment interaction. *Genetics*, **50**, 244.

De Fries, J. C. (1972) Quantitative aspects of genetics and environment in the determination of behavior. In Ehrman, L., Omenn, G. S., and Caspari, W. E. (eds), *Genetics, Environment and Behavior: Implications for Education Policy,* 5–16. New York, Acad. P.

Denenberg, V. H., and Rosenberg, K. M. (1967) Nongenetic transmission of information. *Nature*, **216**, 549–50.

Dickerson, J. W. T., and Dobbing, J. (1967) Parental and postnatal growth and development of the central nervous system of the pig. *Proc. Roy. Soc.*, ser. B. **166**, 384–95.

Dickerson, J. W. T., and McCance, R. A. (1967) Effect of undernutrition on the postnatal development of the brain and cord in pigs. *Proc. Roy. Soc.*, ser. B, **166**, 396–407.

Dickerson, J. W. T., and Walmsley, A. L. (1967) The effect of undernutrition and subsequent rehabilitation on the growth and composition of the central nervous system of the rat. *Brain*, **90**, 897–906.

Dinkel, R. M. (1952) Occupation and fertility in the United States. *Am. Soc. Rev.*, **17**, 178–83.

Dobbing, J., and Widdowson, E. M. (1965) The effect of undernutrition and subsequent rehabilitation on myelination of rat brain as measured by its composition. *Brain*, **88**, 357–66.

Dobzhansky, T. (1955) *Evolution, Genetics and Man*. New York, Wiley.

Dobzhansky, T., and Spassky, B. (1966) Effects of selection and migration on geotactic and phototactic behaviour of drosophila. *J. Roy. Soc.*, ser. B, **168**, 27–47.

Douglas, M. (1966) Population control in primitive groups. *Brit. J. Sociol.*, **17**, 263–73.

Dreger, R. M. and Miller, K. S. (1960) Comparative psychological studies of Negroes and whites in the United States. *Psychol. Bull.*, **57**, 361–402.

Dreger, R. M., and Miller, K. S. (1968) Comparative psychological studies of Negroes and whites in the United States. *Psychol. Bull. Monog. Supp.*, **70** (3), pt 2.

Dubin, J. A., Osburn, H., and Winick, D. M. (1969) Speed and practice: effects on Negro and white test performance. *J. Appl. Psychol.*, **53**, 19–23.

Dubos, R. (1970) *So Human an Animal*. London, Hart-Davis.

Duesenberry, J. S. (1960) *Comment. Demographic and Econ. Change in Developed Countries*, 231–4. Princeton UP.

Dunn, L. C. (1964) Abnormalities associated with a chromosome region in the mouse. *Science*, **144**, 260–3.

Easterlin, R. A. (1969) Towards a socio-economic theory of fertility: a survey of recent research on economic factors in American fertility. In Behrman, S. J., Corsa, L., and Freedman, R. (eds) *Fertility and Family Planning*, 127–56. Ann Arbor, U Michigan P.

Eaves, L. J., and Jinks, J. L. (1972) Insignificance of evidence for differences in heritability of IQ between races and social classes. *Nature*, **240** (5376), 84–8.

Eayrs, J. T., and Horn, G. (1955) Development of cerebral cortex in hypothyroid and starved rats. *Anat. Record*, **121**, 53–61.

Eckland, B. K. (1965) Academic ability, higher education, and occupational mobility. *Am. Soc. Rev.*, **30**, 735–46.

Eckland, B. K. (1967) Genetics and sociology: a reconsideration. *Am. Soc. Rev.*, **32**, 173–94.

Eckland, B. K., and Kent, D. P. (1968) Socialization and social structure. In *Perspectives on Human Deprivation: Biological, Psychological and Sociological*. Bethesda, Md., Nat. Inst. of Child Health and Hum. Devel.

Eibel-Eibesfeldt, I. (1972) *Die Ko Buschmann-Gesellschaft: Gruppenbildung und Aggressions kontrolle*. Munich, Piper.

Eichenwald, H. F., and Fry, P. C. (1969) Nutrition and learning. *Science*, **163**, 644–8.

Engel, R. (1956) Abnormal brain-wave patterns in kwashiorkor. *Electroencephalography and Clin. Neurophysiol.*, **8**, 489–500.

Erlenmeyer-Kimling, L., and Jarvik, L. F. (1963) Genetics and intelligence: a review. *Science*, **142**, 1477–9.

Esselin, L. (1959) Charles Darwin, Edward Blyth, and the theory of natural selection. *Proc. Am. Phil. Soc.*, **103**, 94–158.

Eysenck, H. J. (1967) Intelligence assessment: a theoretical and experimental approach. *Brit. J. Educ. Psychol.,* **37,** 81–98.

Eysenck, H. J. (1971) *Race, Intelligence and Education.* London, Temple-Smith.

Eversley, D. (1959) *Social Theories of Fertility and the Malthusian Debate.* London, Oxford UP.

Firth, R. (1939) *Primitive Polynesian Economy.* London, Routledge & Kegan Paul.

Firth, R. (1957) *We the Tikopia.* London, Allen & Unwin.

Firth, R. (1959) *Social Change in Tikopia.* London, Allen & Unwin.

Fisher, R. A. (1918) The correlation between relatives on the supposition of Mendelian inheritance. *Trans. Roy. Soc. Edin.,* **52,** 399–433.

Fisher, R. A. (1930) *The Genetical Theory of Natural Selection.* Oxford, Clarendon.

Fishman, M. A., Prensky, A. L., and Dodge, P. R. (1969) Low content of cerebral lipids in infants suffering from malnutrition. *Nature,* **221,** 552–3.

Floud, J. E., Halsey, A. H., and Martin, J. M. (1957) *Social Class and Educational Opportunity.* London, Heinemann.

Fricke, H. (1971) Fische als Feinde tropische Seegel. *Marine Biol.,* **9,** 328–38.

Frisch, K. von (1967) *The Dance Language and Orientation of Bees.* Cambridge, Mass., Harvard UP.

Galton, F. (1889) *Natural Inheritance.* London, Macmillan.

Getzels, J. W., and Jackson, P. W. (1962) *Creativity and Intelligence: Explorations with Gifted Students.* New York, Wiley.

Gitmez, A. S. (1971) *Instructions as Determinants of Performance.* Paper read at conference on Cultural Factors in Mental Test Development, Application and Interpretation, sponsored by the NATO Advisory Group on Human Factors and the Turkish Scientific and Technical Research Council, Istanbul.

Glass, D. (ed.) (1954) *Social Mobility in Britain.* London, Routledge & Kegan Paul.

Glass, D. (ed.) (1968) *Genetics.* New York, Rockefeller UP and Russell Sage Foundation.

Goslin, D. (1963) *The Search for Ability.* New York, Russell Sage Foundation.

Gray, S. W., and Klaus, R. A. (1966) *Deprivation, Development and Diffusion.* Presidential address. Am. Psychol. Assoc.

Gray S. W., and Klaus, R. A. (1969) *The Early Training Project: A Seventh Year Report.* Peabody College, Nashville, Demonstration and Research Center for Early Education.

Gruenwald, P. (1968) In Waisman, H. A., and Kerr, G. R. (eds) *Foetal Growth and Development.* New York, McGraw-Hill.

Guthrie, H. A., and Brown, M. L. (1968) Effect of severe undernutrition in early life on growth, brain size and composition in adult rats. *J. Nutrition,* **94,** 419–26.

Haldane, J. B. S. (1932) *The Causes of Evolution.* London, Longmans Green.

Haldane, J. B. S. (1937) The effect of variation on fitness. *Am. Nat.,* **71,** 337–49.

Haldane, J. B. S. (1957) The cost of natural selection. *J. Genet.,* **55,** 511–24.

Hall, V. A., and Turner, R. R. (1974) The validity of the 'different language explanation' for poor scholastic performance by black students. *Rev. Educ. Res.,* **44,** 69–81.

Haller, M. H. (1963) *Eugenics: Hereditarian Attitudes in American Thought.* Rutgers UP.

Halsey, A. H. (1959) Class differences in intelligence. *Brit. J. Statist. Psychol.,* **12,** 1–4.

Halsey, A. H. (1972) *Educational Priority,* vol. 1 : *EPA Problems and Policies.* London, HMSO.

Hardy, A. C. (1960) Was man more aquatic in the past? *New Sci.,* **7,** 642–5.

Harlow, H. F., and Zimmermann, R. R. (1959) Affectional response in the infant monkey. *Science,* **130,** 431.

Harris, H. (1970) *The Principles of Human Biochemical Genetics.* Amsterdam and London, North–Holland.

Harrison, G. A., and Boyce, A. J. (eds) (1972) *The Structure of Human Populations.* Oxford, Clarendon.

Hawthorn, G. P. (1970) *The Sociology of Fertility.* London, Collier-Macmillan.

Hayes, H. K., and Immer, F. R. (1942) *Methods of Plant Breeding.* New York and London, McGraw-Hill.

Hick, W. E. (1952) On the rate of gain of information. *Quart. Exp. Psychol.,* **4,** 11–26.

Higgins, J. V., Reed, E. W., and Reed, S. C. (1962) Intelligence and family size: a paradox resolved. *Eugenics Quart.,* **9,** 2.

Hinde, J. S., and Hinde, R. A. (eds) *Constraints on Learning: Limitations and Predispositions.* London and New York, Acad. P.

Hinde, R. A. (1966) *Animal Behaviour: A Synthesis of Ethology and Comparative Psychology.* New York, McGraw-Hill (2nd edn 1970).

Ho, P. T. (1962) *The Ladder of Success in Imperial China.* New York, Columbia UP.

Holt, J. (1970) *How Children Fail.* Harmondsworth, Penguin Books.

Holt, J. (1971) *How Children Learn.* Harmondsworth, Penguin Books.

Hooton, E. A. (1926) Significance of the term race. *Science,* **63,** 75–81.

Houghton, V. (1966) Intelligence testing of West Indian and English children. *Race* (October).

Humphreys, L. G. (1973) Implications of group differences for test interpretation. *Assessment in a Pluralistic Society: Proc. 1972 Invitational Conf. on Testing Problems,* 56–71. Princeton, N.J., Educational Testing Service.

Hunt, J. (1961) *Intelligence and Experience.* New York, Ronald P.

Husén, T. (1974) *Talent, Equality and Meritocracy.* Amsterdam, European

Hutt, C., and Hutt, S. J. (1970) *Direct Observation and Measurement of Behavior.* Springfield, Ill., Charles C. Thomas.

Hytten, F. E., and Leitch, I. (1971) *The Physiology of Human Pregnancy.* Oxford, London and Edinburgh, Blackwell.

Ilg, F. L., and Ames, L. B. (1964) *School Readiness: Behavior Tests Used at the Gesell Institute.* New York, Harper & Row.

Irvine, S. H. (1969) Contributions of ability and attainment testing in Africa to a general theory of intellect. *J. Biosoc. Sci.,* **1,** suppl. 1.

Jackson, C. M., and Stewart, C. A. (1920) The effects of inanition in the young upon the ultimate size of the body and the various organs in the albino rat. *J. Exp. Zool.,* **30,** 8.

James, T. (1975) *West Riding EPA Project follow-up studies.* Rep. to the SSRC.

Jencks, C. (1972) *Inequality.* New York, Basic Books.

Jensen, A. R. (1969) How much can we boost IQ and scholastic achievement? *Harvard Educ. Rev.,* **39** (1), 1–123.

Jensen, A. R. (1970) Hierarchical theories of mental ability. In Dockrell, B. (ed.) *On Intelligence,* 119–90. Toronto, Ontario Inst. Studies in Educ.

Jensen, A. R. (1971a) A two-factor theory of familial mental retardation. *Human Genetics: Proc. of the 4th Internat. Cong. of Human Genetics.* Amsterdam, Excerpta Medica.

Jensen, A. R. (1971b) Can we and should we study race differences? in Brace, C. L., Gamble, G. R., and Bond, J. T. (eds) *Race and Intelligence.* Anthrop. Studies No. 8. Washington, DC, Am. Anthrop. Assoc.

Jensen, A. R. (1973a) *Educability and Group Differences.* New York, Harper & Row; London, Methuen.

Jensen, A. R. (1973b) The differences are real. *Psychol. Today,* **7,** 80–6.

Jensen, A. R. (1973c) Level I and Level II abilities in three ethnic groups. *Am. Educ. Res. J.,* **10,** 263–76.

Jensen, A. R. (1974a) Cumulative deficit: a testable hypothesis? *Devel. Psychol.,* **10,** 996–1019.

Jensen, A. R. (1974b) Interaction of Level I and Level II abilities with race and socioeconomic status. *J. Educ. Psychol.,* **66,** 99–111.

Jensen, A. R. (1974c) How biased are culture-loaded tests? *Genet. Psychol. Monogs,* **90,** 185–244.

Jensen, A. R. (1974d) The effect of race of examiner on the mental test scores of white and black pupils. *J. Educ. Meas.,* **11,** 1–14.

Jensen, A. R. (1975) A theoretical note on sex linkage and race differences in spatial visualization ability. *Behav. Genet.,* **5** (2).

Jinks, J. L., and Eaves, L. J. (1974) IQ and inequality. *Nature,* 22 March.

Jinks, J. L., and Fulker, D. W. (1970) Comparison of the biometrical, genetical, MAVA and classical approaches to the analysis of human behaviour. *Psychol. Bull.,* **73,** 311–49.

John, V. (1971) Whose is the failure? In Brace, C. L., Gamble, G. R., and Bond, J. T. (eds) *Race and Intelligence.* Anthrop. Studies No. 8. Washington, DC, Am. Anthrop. Assoc.

Juel-Nielsen, N. (1965) Individual and environment: a psychiatric-psychological investigation of monozygous twins reared apart. *Acta Psychiatrica et Neurologica Scandinavia,* Monog. Supp. 183.

Kagan, J. (1968) On cultural deprivation. In Glass, D. G. (ed.) *Environmental Influences: Proceedings of the Conference.* New York, Rockefeller UP and Russell Sage Foundation.

Kamin, L. J. (1973) *Heredity, Intelligence, Politics and Psychology*. Paper presented to 13th Internat. Cong. of Genetics. Berkeley, Calif.

Kanner, L. (1943) Autistic disturbances of affective contact. *Nerv. Child,* **2,** 217–50.

Katz, I., and Benjamin, L. (1960) Effects of white authoritarianism in biracial work groups. *J. Abn. Soc. Psychol.,* **61** (3), 448–56.

Katz, I., and Greenbaum, C. (1963) Effects of anxiety, threat, and racial environment on task performance of Negro college students. *J. Abn. Soc. Psychol.,* **66** (6) 562–7.

Katz, I. *et al.* (1964) The influence of race of the experimenter and instructions upon the expression of hostility by Negro boys. *J. Soc. Issues,* **20** (2), 54–9.

Katz., I., *et al.* (1965) Effects of task difficulty, race of administrator, and instructions on digit-symbol performance of Negroes. *J. Pers. Soc. Psychol.,* **2** (1), 53–9.

Kennedy, W. P. (1967) Epidemiologic aspects of the problem of congenital malformations. *Birth Defects,* orig. ser., **3** (2).

Keyfitz, N. (1972) Population theory and doctrine: a historical survey. In Petersen, W. (ed.) *Readings in Population,* 41–69. New York, Macmillan.

Kimura, M. (1966) Current fertility patterns in Japan. *World Population Conference 1965.* United Nations.

King, J. C. (1971) *The Biology of Race.* New York, Harcourt, Brace & World.

King, J. L. (1967) Continuously distributed factors affecting fitness. *Genetics,* **55,** 483–92.

King, J. L., and Jukes, T. H. (1969) Non-Darwinian evolution. *Science,* **164,** 788–98.

Kirk, D. (1960) *Influence of Business Cycles on Marriage and Birth Rates in Demographic and Economic Changes in Developed Countries,* 241–60. Princeton UP.

Knodel, J. E. (1974) *The Decline of Fertility in Germany 1871–1939.* Princeton UP.

Konishi, M., and Nottebohm, F. (1969) Experimental studies in the ontogeny of avian vocalizations. In Hinde, R. A. (ed.) *Bird Vocalizations in Relation to Current Problems in Biology and Psychology.* Cambridge UP.

Kruuk, H. (1972) *The Spotted Hyena.* Chicago UP.

Kummer, H. (1968) *The Social Organisation of Hamadryas Baboons.* Basel, S. Karger.

Kuznets, S. (1958) *Six Lectures on Economic Growth.* New York, Free Press.

Lavin, D. E. (1965) *The Prediction of Academic Performance.* New York, Russell Sage Foundation.

Lee, R. B. and Devore, I. (eds) (1968) *Man the Hunter.* Chicago, Aldine.

Lehrman, D. S. (1970) Semantic and conceptual issues in the nature–nurture problem, In Aronson, L. R., Tobach, E., Lehrman, D. S., and Rosenblatt, J. S. (eds) *Development and Evolution of Behavior.* San Francisco, Freeman.

Leibenstein, H. (1957) *Economic Backwardness and Economic Growth.* London, Chapman & Hall.

Leibenstein, H. (1975) The economic theory of fertility decline. *Quart. J. Econ.,* **89,** 1–31.

Lesser, G. S., Fifer, G., and Clark, D. H. (1965) Mental abilities of children from different social class and cultural groups. *Monog. Soc. Res. Child Devel.,* **30** (4).

Levins, R. (1968) *Evolution in Changing Environments.* Princeton UP.

Lewontin, R. C. (1967) An estimate of average heterozygosity in man. *Am. J. Hum. Genet.,* **19,** 681–5.

Linn, R. L. (1973) Fair test use in selection. *Rev. Educ. Res.,* **43,** 139–61.

Lipset, S. M., and Bendix, R. (1960) *Social Mobility in Industrial Society.* London, Heinemann.

Little, A., and Smith, G. (1971) *Strategies of Compensation: A Review of Educational Projects for the Disadvantaged in the United States.* Paris, OECD.

Livi-Bacci, M. (1971) *A Century of Portuguese Fertility.* Princeton UP.

Lorenz, K. (1965) *Evolution and Modification of Behavior.* Chicago UP.

Lorenz, K. (1973) *Die Ruckseite des Spiegels: Versuch einer Naturgeschichte menschlichen Erkennens.* Munich, Piper Verlag.

Lush, J. L. (1968) Genetic unknowns and animal breeding a century after Mendel. *Trans. Kansas Acad. Sci.,* **71,** 309–14.

McClearn, G. E., and De Fries, J. C. (1973) *Introduction to Behavioral Genetics.* San Francisco, Freeman.

MacFarlane, A. (1976) *Resources and Population: A Study of the Gurungs of Nepal.* Cambridge UP.

McGrew, W. C. (1972) *An Ethological Study of Children's Behavior.* London and New York, Acad. P.

Malthus, T. R. (1970) *An Essay on the Principle of Population.* Flew, A. (ed.) Harmondsworth, Penguin Books.

Mamdani, M. (1972) *The Myth of Population Control.* New York, Mon. Rev. P.

Mandelbaum, D. G. (1974) *Human Fertility in India.* Berkeley, U California P.

Manning, A. (1956) Some aspects of the foraging behaviour of bumble-bees. *Behaviour,* **9,** 164–203.

Marler, P. (1967) Comparative study of song development in sparrows. *Proc. Internat. Orn. Cong.,* **14,** 231–44.

Mather, K. (1955) Polymorphism as an outcome of disruptive selection. *Evolution,* **9,** 52–61.

Mather, K. (1956) The effect on the distribution of intelligence of increasing the heritable variation. In *The Hazards to Man of Nuclear and Allied Radiations.* Cmd. 9780. London, HMSO

Mayr, E. (1963) *Animal Species and Evolution.* Cambridge, Mass., Belknap.

Mech, L. D. (1966) The wolves of Isle Royale. *U.S. Nat. Park Serv. Fauna,* ser. 7, 1–210.

Mech, L. D. (1970) *The Wolf.* London, Constable.

Mercer, J. (1972) IQ: the lethal label. *Psychol. Today.* **6** (4), 44–7.

Merton, R. K. (1957) *Social Theory and Social Structure.* New York, Free P.

Miller, S. M. (1960) *Comparative Social Mobility.* Oxford, Blackwell.

Mincer, J. (1963) Market prices, opportunity costs and income effects. In Christ, C. F. (ed.) *Measurement in Economics.* Stanford UP.

Morris, D. (1967) *The Naked Ape.* London, Cape.

Mørch, E. T. (1941) Chondrodystrophic dwarfs in Denmark, *Opera ex Domo biol. hered. hum. Univ. Hafniensis,* **3,** 1–200.

Mourant, A. E., Kopec, A. C., and Domaniewska-Sobczak, K. (1976) *The Distribution of Human Blood Groups and Other Polymorphisms.* London, Oxford UP, 2nd Edn.

Moynihan, D. P. (ed.) (1968) *On Understanding Poverty.* New York, Basic Books.

Muller, H. J. (1950) Our load of mutations. *Am. J. Hum. Genet.,* **2,** 111–76.

Muller, H. J. (1963) Genetic progress by voluntarily conducted germinal choice. In Wolstenholme, G. (ed.) *Man and His Nature.* Boston, Mass., Little Brown.

Neel, J. V., and Schull, W. J. (1968) On some trends in understanding the genetics of man. *Perspect. Biol. & Med.,* **11,** 565–60.

Nei, M., and Roychoudhury, A. K. (1973) Gene differences and effective divergence times between the three major races of man: Caucasoid, Negroid and Mongoloid. *Genetics,* **74** (2), 193.

Nelson, G. K. (1959) The electroencephalogram in kwashiorkor. *Electroencephalography and Clin. Neurophysiol.,* **11,** 73–84.

Nelson, G. K. (1963) Electroencephalographic studies in sequelae of kwashiorkor and other diseases in Africans. In Snowball, G. J. (ed.) *Science and Medicine in Central Africa: Proc. Central African Sci. Med. Cong.* Oxford, Pergamon.

Newman, H. H., Freeman, F. N., and Holzinger, K. J. (1937) *Twins: A Study of Heredity and Environment.* Chicago UP.

Newsom, J. (1963) *Half Our Future.* London, HMSO.

Nicholson, A. J. (1954) Compensatory reactions of populations to stresses, and their evolutionary significance. *Aust. J. Zool.,* **2,** 1–8.

Noble, C. E. (1969) Race, reality, and experimental psychology. *Perspect. Biol. & Med.,* **13,** 10–30.

Norton-Griffiths, M. (1967) Some ecological aspects of the feeding. behaviour of the oystercatcher on the edible mussel. *Ibis,* **109,** 412–24.

Office of Population Censuses and Surveys (1974) *The Registrar General's Quarterly Return for England and Wales: Quarter Ended 30 June 1974.* London, HMSO.

O'Gorman, G. (1970) *The Nature of Childhood Autism.* London, Butterworth.

Payne, J. (1974) *Educational Priority,* vol. 2: *Surveys and Statistics.* London, HMSO.

Penrose, L. S. (1948) The supposed threat of declining intelligence. *Am. J. Ment. Def.*

Penrose, L. S. (1949) *The Biology of Mental Defect.* London, Sidgwick & Jackson.

Pettigrew, T. (1964) *A Profile of the Negro American.* Princeton, N.J., Van Nostrand.

Poortinga, Y. (1972) A comparison of African and European students in simple auditory and visual tasks. In Cronbach, L. J., and Drenth, P. J. D. (eds) *Mental Tests and Cultural Adaptation*, 349–54. The Hague, Mouton.

Popham, R. M. (1953) The calculation of reproductive fitness and the mutation rate of the gene for chondrodystrophy. *Am. J. Hum. Genet.*, **5**, 73–5.

Rasmussen, K. (1931) *The Netsilik Eskimos: Report of the Fifth Thule Expedition*, vol. 8.

Reed, T. E. (1969) Caucasian genes in American Negroes. *Science*, **165**, 762–8.

Royal Commission on Population (1949) *Report*. London, HMSO.

Robinson, W., and Horlacher, D. (1971) Population growth and economic welfare. *Reports on Population/Family Planning*, 6.

Ryder, N. B. (1965) The measurement of fertility patterns. In Sheps, M. C., and Ridley, J. C. (eds) *Public Health and Population Change*, 287–309. Pittsburgh UP.

Ryder, N. B., and Westoff, C. F. (1971) *Reproduction in the United States 1965*. Princeton UP.

Salaman, S. (1949) *The History and Social Influence of the Potato*. Cambridge UP.

Scarr-Salapatek, S. (1971) Race, social class and IQ. *Science*, **174**, 1285–95.

Schaller, G. B. (1972a) Predators of the Serengeti. *Nat. Hist.*, **81**, 38–49, 60–9.

Schaller, G. B. (1972b) *The Serengeti Lion*. Chicago UP.

Schaller, G. B., and Lowther, G. R. (1969) The relevance of carnivore behavior to the study of early hominids. *S. West J. Anthrop.*, **25**, 307–41.

Schapiro, S., and Vukovich, K. R. (1970) Early experience effects upon cortical dendrites: a proposed model for development. *Science*, **167**, 992–4.

Schultz, T. W. (ed.) (1974) *Economics of the Family*. Chicago UP.

Schmalhausen, I. I. (1949) *Factors of Evolution*. Philadelphia, Blakiston.

Schwidetzky, I. (1967) Race and the biological history of peoples. In Kuttner, R. E. (ed.) *Race and Modern Science*. New York, Soc. Sci. P.

Scottish Council for Research in Education (1949) *The Trend of Scottish Intelligence*. U London P.

Scottish Council for Research in Education (1953) *Social Implications of the 1947 Scottish Mental Survey*. U London P.

Scrimshaw, N. S., and Gordon, C. E. (eds) (1968) *Malnutrition, Learning and Behaviour*. Cambridge, Mass., MIT P.

Seligman, M. E. P. (1970) On the generality of the laws of learning. *Psychol. Rev.* **77**, 406–18.

Sewell, W. H., and Hauser, R. (1972) Causes and consequences of higher education: models of the status and attainment process. *Am. J. Agric. Econ.*, **54**,

Sewell, W. H., *et al.* (1969) The educational and early occupational attainment process. *Am. Soc. Rev.*, **4**.

Shields, J. (1962) *Monozygotic Twins: Brought up Apart and Brought up Together.* London, Oxford UP.

Shuey, A. M. (1958, 2nd edn 1966) *The Testing of Negro Intelligence.* London, Holborn; New York, Soc. Sci. P.

Simon, J. L. (1974) *The Effects of Income on Fertility.* Chapel Hill, Carolina Pop. Center.

Smith, G. (1975) *Educational Priority,* vol. 4: *The West Riding EPA.* London, HMSO.

Smith, M. (1973) A cross-national and cross-temporal analysis of the cultural component built into the Stanford-Binet Intelligence Scale and other Binet tests. Paper presented to the 13th Internat. Cong. Genetics, Berkeley, Calif. (Published in *Genet. Psychol. Monogs.* (1974), **89,** 2nd half, 307–34.)

Smith, M. S., and Bissell, J. S. (1970)Report analysis: the impact of Head Start. *Harvard Educ. Rev.,* **40** (1), 51–104.

Sorokin, P. (1927) *Social Mobility.* New York, Harper.

Spencer, P. (1965) *The Samburu.* London, Routledge & Kegan Paul.

Spencer-Booth, Y., and Hinde, R. A. (1971) Effects of brief separations from mothers during infancy on behaviour of Rhesus monkeys 6–24 months later. *J. Child Psychol. Psychiat.,* **12,** 157–72.

Spuhler, J. N. (1963) The scope for natural selection in Man. In Schull, W. J. (ed.) *Genetic Selection in Man.* Ann Arbor, U Michigan P.

Stallings, J. (1974) *Follow-through Classroom Observation Evaluation, 1972–3.* Stanford Res. Inst.

Stallings, J. (1975) *Relationships between Classroom Instructional Practices and Child Development.* Stanford Res. Inst.

Stebbins, G. L. (1963) Perspectives, 1. *Am. Sci.,* **51,** 362.

Stevenson, A. C. (1959) The load of hereditary defect in human populations. *Radiat. Res. Suppl.,* **1,** 306–25.

Stevenson, A. C. (1961) Frequency of congenital and hereditary disease. *Brit. Med. Bull.,* **17,** 254–9.

Steward, J. (1938) The Basin Plateau Indians. *Smithsonian Inst. Bur. of Am. Ethnol. Bull.,* **120.**

Stoch, M. B., and Smythe, P. M. (1963) Does undernutrition during infancy inhibit brain growth and subsequent intellectual development? *Archives of Disease in Childhood,* **38,** 546–52.

Stoch, M. B., and Smythe, P. M. (1967) The effect of undernutrition during infancy on subsequent brain growth and intellectual development. *S. Afr. Med. J.,* **41,** 1027–31.

Strøm, A. (ed.) (1968) *Norwegian Concentration Camp Survivors.* Oslo, Universitetsforlaget.

Sved, J. A., Reed, T. E., and Bodmer, W. F. (1967) The number of balanced polymorphisms that can be maintained in a natural population. *Genetics,* **55,** 469–81.

Tabbarah, R. B. (1971) Toward a theory of demographic development. *Econ. Devel. & Cult. Change,* **19,** 257–76.

Teissier, G. (1945) Mecanisme de l'evolution. *La Pensée,* **2,** 5–19.

Thoday, J. M. (1953) Components of fitness. *Symp. Soc. Exp. Biol.,* **7,** 96–113.

Thoday, J. M. (1959) Effects of disruptive selection, I: genetic flexibility. *Heredity*, **13**, 187–203.

Thoday, J. M. (1963) Causes and functions of genetic variety. *Eugenics Rev.*, **53**, 4–26.

Thoday, J. M. (1969) Limitations to genetic comparisons of populations. *J. Biosoc. Sci.*, 1, suppl. 1. In Harrison, G. A., and Peel, J. (eds) *Biosocial Aspects of Race: Proc. 5th Ann. Symp. Eugenics Soc.*, 3–14. London (1968).

Thorpe, W. H. (1961) *Bird Song*. Cambridge UP.

Tiger, L. (1969) *Men in Groups*. New York, Random House.

Tinbergen, E. A., and Tinbergen, N. (1972) Early childhood autism: an ethological approach. *Z. Tierpsychol.*, suppl. 10, 1–53. Berlin, Parey.

Tinbergen, N. (1958) *Curious Naturalists*. London, Country Life.

Tinbergen, N. (1972) *The Animal in its World*, vol. 1: *Field Studies*. London, Allen & Unwin.

Tizard, B. (1974) *Preschool Education in Great Britain: A Research Review*. Soc. Sci. Res. Council.

Tobias, P. V. (1970a) Puberty, growth, malnutrition, and the weaker sex: and two new measures of environmental betterment. *The Leech*, **51** (4), 101–7.

Tobias, P. V. (1970b) Brain size, grey matter and race: fact or fiction? *Am. J. Phys. Anthrop.*, new ser., **32** (1), 3–26.

Tobias, P. V. (1972a) Growth and stature in Southern African populations. In Vorster, D. J. (ed.) *The Human Biology of Environmental Change: Proc. IBP Adaptability Conf., Blantyre, Malawi* (1971).

Tobias, P. V. (1972b) *The Meaning of Race*. Johannesburg, S. African Inst. Race Relations, 2nd rev. enl. edn.

Toki, E. (1971) *Kindergarten Children's Preferences for Color, Form, Size and Number Stimuli*. Unpub. Master's thesis, U California, Berkeley.

Tuddenham, R. D. (1948) Soldier intelligence in World Wars I and II. *Am. Psychol.*, **3**, 54–6.

Tuddenham, R. D. (1970) A 'Piagetian' test of cognitive development. In Dockerell, B. (ed.) *On Intelligence*. Toronto, Ontario Inst. Studies in Educ.

Turner, V. W. (1957) *Schism and Continuity*. Manchester UP.

United Nations Dept of Social Affairs, Population Division (1953) *The Determinants and Consequences of Population Trends*. ST/SOA ser. A, Pop. Stud. No. 17. New York.

Urbach, P. (1974) Progress and degeneration in the IQ debate. *Brit. Phil. Sci.*, **25**, 99–135, 235–59.

Vernon, P. E. (1955) The assessment of children. *Studies in Educ.*, No. 7. U. London Inst. Educ.

Wallace, A. R. (1869) *Contributions to the Theory of Natural Selection*. London.

Wallace, B. (1968) *Topics in Population Genetics*. New York, Norton.

Wallace, B. (1970) *Genetic Load*. Engelwood Cliffs, NJ, Prentice-Hall.

Washburn, S. L. (1968) Behaviour and the origin of man. *Rockefeller Univ. Rev.* (January), 10–19.

Washburn, S. L., and Lancaster, C. S. (1968) The evolution of hunting. In Lee, R. B. and Devore, I. (eds) *Man the Hunter*, 293–304. Chicago, Aldine.

Watley, D. J. (1969) Career or marriage? A longitudinal study of able young women. *Nat. Merit Scholarship Rep.*, 5.

Watson, P. (1972) IQ: the racial gap. *Psychol. Today*, 6 (4), 48–50, 97.

Weber, M. (1947) *The Theory of Social and Economic Organization*. London, Oxford UP.

Wehmer, F. R. H., and Scales, B. (1970) Prenatal stress influences the behaviour of subsequent generations. *Communications in Behav. Biol.*, 5, 211–14.

Weikart, D. P. (1967) *Results of Preschool Intervention Programs*. Ypsilanti, Michigan.

Weikart, D. P. (1972) Relationship of curriculum, teaching and learning in preschool education. In Stanley, J. C. (ed.) *Preschool Programs for the Disadvantaged*, 22–6. Baltimore, Md, Johns Hopkins.

Westoff, C. F. (1966) Fertility control in the United States. In *World Population Conf. 1965*. New York, United Nations.

Whelpton, P. K., Campbell, A. A., and Patterson, J. E. (1966) *Fertility and Family Planning in the United States*. Princeton UP.

Wickler, W. (1967) Socio-sexual signals and their intra-specific imitation among primates. In Morris, D. (ed.) *Primate Ethology*, 69–148. London, Weidenfeld & Nicolson.

Wiener, N. (1948) *Cybernetics*. New York, Wiley.

Willmott, P. and Young, M. (1960) *Family and Class in a London Suburb*. London, Routledge & Kegan Paul.

Wilson, A. B. (1967) Educational consequences of segregation in a California community. In *Racial Isolation in the Public Schools*, Appendices, vol. 2 of a report by the US Commission on Civil Rights. Washington, DC, US Govt Printing Office.

Wolff, M., and Stein, A. (1967) *Six Months Later: A Comparison of Children who had Head Start, Summer 1965, with their Classmates in Kindergarten*. New York, Center for Urban Educ.

Wrigley, E. A. (1969) *Population and History*. London, Weidenfeld & Nicolson.

Wynne-Edwards, V. C. (1962) *Animal Dispersion in Relation to Social Behaviour*. Edinburgh, Oliver & Boyd.

Wynne-Edwards, V. C. (1963) Intergroup selection in the evolution of social systems. *Nature*, 200, 623–6.

Wyon, J. B., and Gordon, J. E. (1971) *The Khanna Study*. Cambridge, Mass., Harvard UP.

Yalman, N. (1963) Female purity in Ceylon and southern India. *J. Roy. Anthrop. Inst.*, 93, 25–58.

Zamenhof, S., Van Marthens, E., and Margolis, F. L. (1968) DNA (cell number) and protein in neonatal brain: alteration by maternal dietary protein restriction. *Science*, 160, 322–3.